Learning to Smell

LEARNING to SMELL

Olfactory Perception

from

Neurobiology to Behavior

Donald A. Wilson &

Richard J. Stevenson

THE JOHNS HOPKINS UNIVERSITY PRESS
Baltimore

The Johns Hopkins University Press
2715 North Charles Street
Baltimore, Maryland 21218-4363
www.press.jhu.edu

Library of Congress Cataloging-in-Publication Data
Wilson, Donald A.
    Learning to smell : olfactory perception from neurobiology to behavior /
Donald A. Wilson and Richard J. Stevenson.
        p. cm.
Includes bibliographical references and index.
ISBN 0-8018-8368-7 (hardcover : alk. paper)
1. Smell. I. Wilson, Donald A. II. Stevenson, Richard J. III. Title.
QP458.W52 2006
612.8′6 — dc22   2005027714

A catalog record for this book is available from the British Library.

Contents

This book represents a rather unusual collaboration between a neurobiologist and a psychologist that grew out of the similarities emerging between physiological and psychophysical research in olfaction. The major problem in olfactory behavioral neuroscience is to determine how the brain discriminates one smell, or perceptual odor, from another. Olfactory and chemical sensory systems were one of the first sensory systems to evolve, and some form of chemosensory system is expressed in every living organism from bacteria to primates. Critical aspects of olfactory system anatomy appear highly convergent across both invertebrates and vertebrates, perhaps in evidence of the unique requirements of a system for dealing with complex, often unpredictable stimuli.

The traditional approach for understanding olfactory perception involves identifying how particular features of a chemical stimulus are represented in the olfactory system. This perspective is at odds, however, with a growing body of evidence, from both neurobiology and psychology, which places primary emphasis on synthetic processing and experiential factors — perceptual learning — rather than on the structural features of the stimulus as critical for odor discrimination. Research from our laboratory and others increasingly argues that experience-based, synthetic olfactory processing leads to treatment of multifeature odorants as individual "odor-objects." This is a process similar to, and perhaps evolutionarily predating, visual object perception, a comparison we frequently make in this book. Furthermore, experience-based, synthetic olfactory processing can be multidimensional, with odor representations coming to integrally include, for example, both multimodal components (e.g., taste) and affective components.

In this volume, we present a new theoretical view of olfactory perception that puts old psychological, ethological, and sensory physiological data in a new light and is backed by new psychophysical and physiological data. In the opening chapter we explore the function of olfactory systems in humans and other animals and the unique difficulties presented by highly complex and often unpredictable chemical stimuli. In the second chapter, we discuss the conceptual and historical roots of the stimulus feature extraction/feature detection approach and its detrimental consequences in shaping thinking about olfaction. We then compare this with contemporary thinking about visual and haptic object recognition and how synthetic and experience-based processing incorporated into those fields offers a better model for understanding olfaction. In chapter 3, we review the anatomy of the olfactory system from a novel theoretical view emphasizing function and how known circuitry may allow the type of synthetic information processing we have proposed. Specific comparisons are made with the circuitry underlying visual object perception. Chapter 4 deals with detection and intensity and demonstrates that, even here, experience plays a role in what would typically be considered low-level processes. Chapters 5 and 6 address the nature of odor perceptual quality in animals and humans and present the argument that odor quality is not dictated solely by the physicochemical stimulus; rather, it is a synthetic construct of physicochemical stimulus properties, memory, and biological constraints. This theoretical view not only encompasses a large range of physiological and behavioral findings but also explains some apparent anomalies, such as difficulty in identifying individual odors in mixtures. Chapter 7 explores the large literature on explicit, associative memory for odors in humans and animals and suggests how these findings may be accounted for by the perceptual memory system dealt with in earlier chapters. Finally, in chapter 8 we summarize the evidence favoring an experiential approach to olfactory perception. We then compare the neural and psychological processing that underpins object perception in olfaction and vision and consider the strengths and weaknesses of drawing such an analogy. In addition, we identify existing data that need to be reinterpreted in light of this new view of olfaction, as well as new questions raised by the mnemonic view of odor perception.

In summary, we propose that experience and cortical plasticity play a critical, defining role in odor perception and that current views of a highly analytical, "receptor-centric" process are insufficient to account for current data.

DON WILSON THANKS HIS FAMILY FOR THEIR LOVE AND SUPPORT over the years, his students for their hard work, and the National Institutes of Health and the National Science Foundation for their generous financial support. Dick Stevenson thanks Caroline, Lucy, Harry, Gemma, and Chris and Mike Thomas for their patience and support. Many colleagues played an important role in developing the ideas described here, notably Trevor Case and Bob Boakes. Much of the work reported here (Stevenson) was supported by funding from the Australian Research Council. Both authors thank Vincent Burke and Nancy Wachter at the Johns Hopkins University Press for helping make this project a success.

Learning to Smell

The Function of the Olfactory System
in Animals and Humans

Smell this book. Seriously, close your eyes, bring the pages of the book to your nose and inhale.* Volatile molecules from the pages, the binding glue, the cover, and, depending on the age of the book at this reading, the accumulated dust and debris of storage enter your nose (via the external nares) and pass over your olfactory receptor sheet on their way to your trachea and lungs (via the internal nares). What happens at the olfactory receptor sheet is nothing short of a remarkable bit of analytical chemistry. Current views of peripheral odorant transduction suggest that individual molecular and submolecular features of the myriad different molecular species you just inhaled are each recognized by a small subset out of hundreds of different olfactory receptors. Thus, no single olfactory receptor for book odor exists; rather, the olfactory receptor sheet performs an analysis to identify the scores or hundreds of individual volatile components over the book. At a conscious, perceptual level, however, we have no direct access to the results of

---

* Did you inhale with a single deep fill of the lungs or did you use a series of short, fast sniffs? Either process draws odorant molecules over the olfactory receptor sheet, although short, fast sniffs may allow greater, or at least differential, access of odorants to the recesses within the olfactory epithelium. How you inhale could thus impact the nature of odor perception—a topic we will not address here, but that is receiving increasing attention both in terms of receptor sheet stimulation and central olfactory pathway processing.

this phenomenal peripheral analysis. Instead, we perceive a wholistic, unitary percept of book odor—largely a single perceptual odor object. The perception may have, at most, two to three identifiable major components, but the vast majority of the exquisite analysis occurring at the periphery is beyond our conscious reach.

The focus of this book is the nature and consequences of odor object perception. How odor objects are formed, how they are shaped by experience, and how the process of synthetic object perception results in both unique capabilities and distinct limitations for olfactory perception are some of the questions to be addressed. Exciting new findings in the psychophysics, neurophysiology, and functional anatomy of the olfactory system are brought together to support our thesis (Wilson and Stevenson 2003). Along the way, we will attempt to highlight new avenues of research to which this view of olfaction draws attention.

A fundamental premise here is that odor objects are learned through experience. Odorants and odorant features that co-occur are synthesized through plasticity within central circuits to form single perceptual outcomes that are resistant to background interference, intensity fluctuations, or partial degradation. Learned odor objects may include multimodal components, and recognition of familiar odor objects can be shaped by context, attention, and expectation.

Experience-dependent odor object perception and synthetic odor coding are certainly not new ideas. In 1890, William James, a founding father of modern psychology noted that "every perception is an acquired perception" (James 1890, 78). Specifically addressing olfaction, he made the following observation: "We know that a weak smell or taste may be very diversely interpreted by us, and that the same sensation will now be named as one thing and the next moment as another . . . In this wise one may make a person taste or smell what one will, if one only makes sure that he shall conceive of beforehand as we wish by saying to him 'Doesn't it smell just like, etc.?' (James 1890, 97–98).*

Together, these two quotes from 1890 suggest that simply knowing what molecular features are present at the olfactory receptor sheet is not sufficient to allow accurate prediction of the resulting olfactory percept. Rather, how the activity within the sensory afferent is read and formed into a percept depends on past experience and current expectations, among other things.

* James's use of the term taste here actually implies "flavor," which is intimately dependent on olfaction.

More recently, in 1973 Ernest Polak outlined a model of olfaction wherein he hypothesized a large set of diversely tuned receptor neurons signaling the unique pattern of odor features present in a stimulus. Once the unique pattern of features is extracted, "the brain attempts to recognize this odor image by scanning and resolving it into previously stored patterns" (Polak 1973, 469). Again, the notion of synthesizing odorant features into perceptual wholes based on their similarity to previously learned patterns is suggested. Other important contributors to this line of thought include Robin Hudson and Gordon Shepherd, who have independently outlined the importance of learning in odor discrimination (Hudson 1999) and noted the similarity of odor perception to object perception in vision (Shepherd 2004).

This line of thought on odor perception as an experience-based, synthetic object process, however, threads through a much larger fabric of work emphasizing the analytical events occurring at the early stages of the olfactory pathway. In many ways this emphasis on odorant analysis has been accelerated by the discovery of the large gene family encoding odorant receptors—one of the most important discoveries in the history of the study of ol-

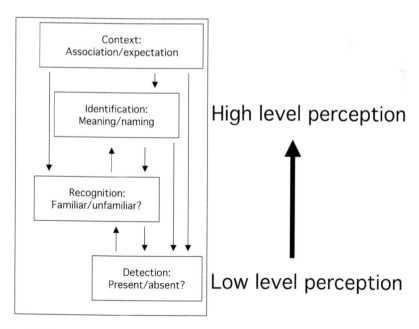

Fig. 1.1. Detection is degree of presence, recognition involves matching input with a stored engram, and identification is the assignment of meaning and naming (in humans). These functions do not necessarily correlate with specific anatomical locations.

faction (Buck and Axel 1991). The exciting finding of this gene family and the resulting powerful tools for analysis of receptor neuron structure and function has led some to conclude (or imply) that smell happens at the receptor sheet and that knowing the pattern of receptor afferent activity will predict the percept (e.g., Mombaerts 2001). However, to quote William James yet again: "Whilst part of what we perceive comes through our senses from the object before us, another part (and it may be the larger part) always comes out of our own head" (James 1890, 103). Thus, to completely understand olfaction, we must not only follow our nose but also use our head!

WHAT IS OLFACTION?

In many ways, olfaction is probably the first sense. Chemosensory systems are expressed in every living animal known, and chemosensitivity emerges very early in ontogeny. Even the vertebrate cranial nerve mediating olfaction is number 1! The importance of chemical stimuli for everything including detection and discrimination of conspecifics, mates, mothers, home, predators, prey, and food places olfaction as a core information-processing system critical for survival and reproductive success in wide-ranging, if not all, animal species. Although humans appear less directly driven by olfactory input than many other animals, odors play an important modulatory role in human attraction, mood, dietary preferences, and detection of danger.

In nature, odorant sources are both environmental (e.g., water, soil, sulfurous minerals) and biological. Biologically derived odors can either be intentional (e.g., intraspecific pheromones or interspecific allelochemicals) or incidental (the odor of bacterial waste in your shoes or floral scents perceived by non-nectar-seeking animals like humans). Given the diversity of situations in which odor cues are used to provide relatively specific information, a basic task of the olfactory system is to allow an animal to answer the question "what is that smell?"

At first glance, this is a relatively straightforward task for the olfactory system. Olfaction mediates sensitivity and perception of volatile chemicals. Chemical stimuli can convey information to the receiver through variation in the precise structure or other properties of the individual odorant molecules, through the concentration of those molecules, and/or through the specific combination and relative concentrations of components in mixtures of odorant molecules. "What is that smell?" could thus simply represent an analytical chemistry problem. Identifying membrane receptors for specific odorant molecular ligands and determining which of those receptors were activated by a particular stimulus should allow us to determine what we are smelling.

In fact, in some specialized cases, simple analytical chemistry may be sufficient to account for olfactory information processing. Infinitesimally small quantities of the pheromone molecule bombykol (E-10, Z-12-hexadecandien-1-ol) signals to the male silkworm moth that a receptive female is nearby and may present an opportunity for mating. Thus, this single molecular species provides significant information to the male moth. By having selective receptors for bombykol and a hardwired central neural circuit, the moth can identify what the smell is and act accordingly.

Using single molecules to signal specific information content can be somewhat limiting, however. To maximize distinctiveness (and thus information content) between different odors used as pheromones or allelochemicals, the underlying volatile molecules must become more complex, with more distinct combinations of specific features. If this process were to rely solely on single molecules as the carriers of information, limits would be reached rapidly. Biological synthesis of complex and larger molecules becomes energetically expensive, and large molecules have reduced volatility and thus reduced transmission speed and distance.

In contrast, use of simple mixtures of small molecules can significantly increase the flexibility of odorant synthesis and signal diversity and thus expands the information content of odors. Small molecules are relatively easy to biosynthesize and tend to be more volatile than larger molecules, enhancing transmission distance. Furthermore, just as combining the simple, 26 units of the alphabet in different mixtures (words) vastly increases information transfer over that possible with the 26 original units, combining a few simple molecules into different mixtures can enhance information transfer through olfaction.

However, as olfactory information processing moves from a situation of hardwired one molecule-one receptor-one meaning toward information processing based on mixtures, new issues arise. First, having a single receptor that selectively binds a specific mixture of molecules rather than a single molecule becomes problematic. Mixtures of molecules can interact in agonistic and antagonistic ways to influence binding at a receptor, but this is not an ideal method for precisely identifying the content of a mixture. Second, even if the system has different receptors to identify the different components of a mixture there must be at least one additional level of processing. For example, discriminating the word lab from ball not only requires three feature detectors for a, b, and l but also a way of monitoring ratios of features, i.e., there are two l's in one stimulus and only one in the other. We know from work on insect pheromones that mixture ratios are incredibly precisely monitored. In oriental fruit moths, even if the pheromonal mixture is composed of 90% of one compound and only 10% of another, the dominant component is not sufficient alone to evoke attraction behavior, rather the complete mixture must be present (Linn et al. 1987). This suggests a form of synthetic processing, wherein the whole is different than the sum of the parts.

Synthetic processing of biologically stable mixtures such as pheromones could occur in a hardwired manner. However, the third problem faced by the olfactory system in dealing with mixtures is that most of the odor mixtures we smell and recognize (e.g., food, home, mate) are not stable over evolutionary time but rather are initially sensed as novel. Thus, the human olfactory system is not hardwired from birth to smell and recognize the odor of coffee or seaweed or eucalyptus or camembert cheese. Each of these stimuli are composed from scores to hundreds of molecular component features, and yet, given some brief familiarity, they are each perceived synthetically as a single-odor object.

The ability to learn new odor objects from previously novel mixtures of components is in some ways the conceptual equivalent of learning to read. Combining a relatively small set of features (e.g., letters) in multiple, different combinations can result in a huge vocabulary of different words. The result is a massive increase in access to information. In evidence of this information expansion, humans, with roughly three hundred different olfactory receptor genes, can discriminate somewhere near ten thousand different odors. Furthermore, by comparing a given sensory input with previously remembered patterns (words), that information can be recognized even if the stimulus changes in **INTENSITY** or is set against a background (as most odors are). In other ways, we and others have argued that learning odor objects is distinctly different than learning to read for the basic reason that synthetic odor objects are highly resistant to perceptual analysis. Thus, visual examination of the word *coffee* leads to the simple analysis that the word contains the components *c, e, f,* and *o*. In contrast, we have virtually no access to the approximately six hundred different volatile components that make up the perceptual odor object coffee. Learned, synthetic odor objects, therefore, allow rapid, robust perception of remarkably complex stimuli, though with limited analysis of their individual component features.

To be useful information in most situations outside of a laboratory, "what is that smell" also implies "and what does that smell mean to me?" If a rat identifies an odorant as trimethylthiazoline (TMT) but does not also rapidly identify TMT as a natural component of fox odor, the rat may be in serious trouble. Yet laboratory rats, inbred and out of contact with foxes for many generations, display selective fear responses to TMT, suggesting that the perception of TMT includes an innate, hardwired meaning to the rat. In fact, the perception of TMT to the rat seems to be inseparable from the meaning—TMT is a fear- and stress-inducing odor stimulus.

Thus, learned perceptual odor objects can also be multifaceted, including meaning or even multimodal components. Odor perceptions can be sickening or sweet or integrally tied to a variety of emotional memories. Just as odor mixtures can be only minimally analyzed into components, multimodal odor perceptual experiences appear inextricably entwined. Our experience of flavor, a mixture of gustation and olfaction, serves as a good example of this perceptual unity. After repeated exposures of a specific odor with a specific taste (e.g., cherry odor with sucrose taste) the perceptual qualities of cherry odor—a purely olfactory stimulus—come to include a sweet (gustatory) component (Stevenson et al. 1995). In fact, the vocabulary of olfaction almost invariably ties the odor to its physical source, e.g., orange or coffee or cheese odors. This is distinctly different than, for example, the vocabulary for color, in which blue, yellow, and red can be distinct percepts in themselves, separate from whatever object produces those reflected wavelengths. Odor mixture components, along with co-occurring sensory inputs and ongoing behavioral states, can become locked into unique perceptual objects through experience.

As noted above, the synthesis of multiple (potentially hundreds) volatile molecules into a single perceptual object can enhance perceptual stability if newly sampled odors are compared with previously stored odor object templates. Computational models of circuitry within the olfactory cortex, for example, suggests a robust ability to complete degraded inputs or recognize familiar patterns of input despite slight disturbances, for example, due to changes in intensity or background odors. This process of storing templates of synthesized odor objects, however, also facilitates discrimination of one odor from another. Given that most odors are complex mixtures of many components, discrimination of one odor from another becomes an incredibly difficult task if individual features, rather than the patterns as a whole, must be compared. Are "ycopaiedenlc" and "ycopciedenlc" the same or different? What about "encyclopedia" and "encyclopedic?" Even though both problems involve the same features, the latter should be easier to solve because of existing internal templates stored in your memory. The same holds for odor discrimination. Familiarity enhances odor discrimination in both humans and animals. We argue that this olfactory perceptual learning occurs because co-occurring odorant features become synthesized through experience-induced cortical plasticity. Upon subsequent exposure to familiar odorants, these learned patterns are evoked and the odor object is recognized.

Learned odor objects of this sort also allow for two other characteristics of odor perception. First, given the multimodal nature of odor objects discussed above, context and expectancy can contribute to the input pattern during the template-matching stage, and in some cases (e.g., James' discussion of weak stimuli above) shape the perceptual outcome in a manner inconsistent with the actual sensory input.

A second consequence of learned odor object perception is an enhanced ability to distinguish odors as separate from background. Odors most likely are never sensed in the absence of background odors, except perhaps within a well-controlled laboratory (even then, it is probably rare). Any given inhalation, therefore, will include odorant features that belong to a target odor and odorant features that belong to the background. The olfactory system seems very capable of filtering out background odors while leaving responsiveness to novel odors intact. We believe this filtering and figure-ground separation is facilitated by a combination of central adaptation (more so than peripheral receptor adaptation) and synthetic odor object processing. Thus, background, stable odorant features are adapted out of the scene, whereas dynamic features are matched to existing templates for object recognition.

EXPERIENCE-DEPENDENT ODOR OBJECT PERCEPTION

The traditional approach for understanding olfactory perception involves identifying how particular features of a chemical stimulus are represented in the olfactory system. However, this perspective is at odds with a growing body of evidence from both neurobiology and psychology, which places primary emphasis on synthetic processing and experiential factors—perceptual learning—rather than on the structural features of the stimulus as critical for odor discrimination. Experience-based, synthetic olfactory processing leads to treatment of multifeature odorants as individual odor objects, similar to, and perhaps evolutionarily predating, visual object perception such as face recognition. Furthermore, experience-based, synthetic olfactory processing can be multidimensional, with odor representations coming to integrally include, for example, both multimodal components (e.g., taste) and affective components. This novel view accounts better for existing data than the traditional view does and comes at a time when the traditional approach is becoming so (prematurely) ingrained that it has reached the status of an accepted truth rather than a tentative statement, as it should be.

In this volume, we present a new theoretical view of olfactory perception that puts old psychological, ethological, and sensory physiological data in a new light; this new view is backed by the latest psychophysical and physiological data. We argue that initial odorant feature extraction/analytical processing is not behaviorally or consciously accessible; rather, it is a first necessary stage for subsequent synthetic processing which in turn drives olfactory behavior. Thus, we propose that experience and cortical plasticity are not only important for traditional associative olfactory memory but also play a critical, defining role in odor perception and that current views of a highly analytical, receptor-centric process are insufficient to account for current data.

The following chapters will describe the historical basis of the analytic and synthetic debate over olfactory processing. Next we will provide an overview of the neurobiology of the olfactory system, highlighting recent findings that emphasize a highly plastic, synthetic neural circuitry. We will then place olfaction in its ecological and ethological context to help understand the evolutionary forces shaping olfactory processing and odor perception. With this background into both the neurobiological tools available for odor processing and the biological and behavioral demands placed on olfaction, the last half of the book lays out our hypotheses of memory-based, synthetic odor object perception. We describe both the advantages and limitations this form of processing places on the olfactory system and odor perception, and highlight new testable predictions stemming from this view.

2

A Historical and Comparative Perspective on Theoretical Approaches to Olfaction

The search for systematic relationships between the physical characteristics of a chemical stimulus and the percept that results from smelling it has been pursued with two quite different goals in mind. The first is that of being able to identify ingredients for use in the flavor and fragrance industry. This approach focuses on understanding how natural odorants come to smell as they do, identifying their components, and then synthesizing chemical analogues to improve availability and consistency. A major motivation has been to reduce cost—a real necessity given that some odorants, such as jasmine, for example, require the collection of 5 million blossoms to make one kilo of essential oil. Contrast this with the cost of synthesizing jasmone or methyl jasmonate, both of which have a jasmine odor and which retail for about six hundred U.S. dollars for 500 kilos (Rossiter 1996). In sum, this is the approach of the organic chemist, and there is a large and well-developed literature detailing the chemical characteristics that correlate with certain types of olfactory sensation and of methods of synthesizing chemicals based on this knowledge (e.g., Beets 1978; Ohloff et al. 1991; Chastrette 1997). Undoubtedly, as this example of jasmine illustrates, the goal of identifying important psychological correlates of an odorant's structure has been both successful and lucrative.

The second goal, and the one that is of principal concern here, is the belief that identifying the relationship between chemical structure and olfactory qualia (and/or behavior) will ultimately allow us to understand the basis of olfactory perception. Not only does the brief historical review below illustrate what a dismal failure this approach has been, but the theory upon which it is based can not deal with the accumulated mass of empirical evidence either—much of which is presented in this book. Before detailing some of the basis for this conclusion, in this case the historical background, it is crucial to bear in mind that receptor-stimulus interactions *must* underpin olfactory perception, just as light falling on the retina *must* underpin vision. The failure lies in assuming that this alone can explain how we smell or see. As we argue below, confusion arises from misinterpreting the correlation between the physicochemical or other such properties of stimuli and their apparent behavioral or experiential consequences. As most statistically literate readers will be aware, correlations cannot inform us about causality.

HISTORICAL VIEW OF OLFACTION AS A PHYSICOCHEMICAL SYSTEM

A stimulus-based theory of olfactory perception is premised upon finding systematic relationships between the stimulus and the resultant odor percept, and we shall refer to this as the basic stimulus-response model of olfaction. Note, however, that few authors historically have been explicit about their stimulus-response models. Rather it was so "obvious" that this approach underpinned olfaction, few then or since felt it necessary to lay out their arguments in detail. In addition, the thinking has changed over time with the advent of new data, so it may seem in our discussion that we have a shifting criterion for what constitutes a stimulus-response model. This is not strictly true. Instead, all we have done is to look at the implications that flow from the various stimulus-response models implied by different workers and to contrast these both with current data and an ecological perspective of olfactory functioning. We should also add that this historical perspective has, as its primary perspective, a psychological orientation. Because advances in experimental psychology have tended in the past to set the pace for sensory physiologists (e.g., Herring on color vision), this seems to be a reasonable approach. In fact one might argue that the current focus on receptor-centric mechanisms owes much to the historical failure of experimental psychology to advance our understanding of olfactory perception. This disconnect that currently exists in olfaction between sensory physiology and experimental psychology is not new. The history of sensory physiological thought on olfaction has followed a course quite distinct from psychological thought and, at times, the two have been polar opposites (see below). One of the goals of this book is to bring these two views into line.

Historically, the search for stimulus-response relationships has been useful because it offered a potential way to discover the number of receptor types that might underpin olfactory perception. One place where one might start to search for such relationships in experimental psychology is in classifying people's descriptions of odors. If relatively few receptors underpinned olfactory perception, then it would be reasonable to assume that some sort of order would emerge when olfactory sensations were described, just as it was possible with descriptions of color or pitch sensation. Once the number of receptor types has then been established, it should be possible to characterize the molecular features needed to selectively stimulate them and to thus produce any olfactory sensation one might desire. This can be seen as analogous to manipulating the frequency of light and sound, with its largely predictable effects on color and pitch sensation, respectively.

Attempts to impose order on the apparently large range of olfactory sensations has a long history in psychology, dating back from Zwaardemaker via Linnaeus to Aristotle. These early schemes were based on individual introspection and, at least in the case of Linnaeus, they were far more concerned with his categorization of flora than with any attempt to understand olfaction. The first attempt to produce an empirical system of classification was that of Hans Henning (1916) and his odor prism (fig. 2.1). Using 415 different odorants, Henning asked his six principal participants to introspect and then identify the "bare sensory quality" of each stimulus. Based on the similarities observed between different odors, he assembled "salient odors" which appeared to form points on an imaginary surface upon which all odors could be located. After trying many arrangements, Henning hit upon the prism design. Not only did he claim that all possible odor sensations could be located on the surface of the prism, he was also quite explicit in stating the stimulus properties that he claimed underpinned his system of classification. Accordingly he developed what appears to be the first physicochemical model, which has features of the stimulus specifically related to the qualitative parameters of the prism.

Henning's book *Der Geruch* (Smell) was especially important, not so much because it foreshadowed decades of work along the same lines that it did, but because it made testable claims. Two examples should suffice to illustrate the problems that arose when the prism's implications were tested.

First, it was found that certain odors generated sensations that the geometry of the prism could not accommodate, that is they fell within the prism, rather than upon its surface (e.g., Hazzard 1930). Drawing even this conclusion was difficult because of poor interrater reliability (e.g., Findley 1924). Second, attempts to test Henning's predicted relationships between the physicochemical characteristics of a stimulus and odor quality were not confirmed either. Chemicals with a known structure that should, according to his scheme, have generated sensations falling on a particular part of the prism, often did not (MacDonald 1922). Although Henning's scheme was largely disconfirmed, it cast a long shadow.

Crocker and Henderson (1927) felt that there were too many primary qualities in Henning's (1916) classification system. They argued that all odor sensations could be captured by the use of just four qualities, fragrant (e.g., benzyl acetate), burnt (e.g., guaiacol), acid (e.g., acetic acid), and caprylic (e.g., butyric acid). Crocker and Henderson (1927) were also quite specific in proposing that these four basic sensations could capture any particular odor quality through each odor's resemblance to these four basic sensations. An interesting aspect to emerge from this work was that it involved, for the first time, explicit similarity (or resemblance) judgments between the to-be-

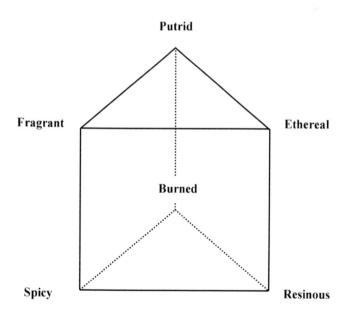

Fig. 2.1. Henning's odor prism.

classified odor and their four standard odors, which could either be physically present or represented by the four labels described above (for later application of this idea, see, for example, Schutz 1964). More importantly, Crocker and Henderson (1927) were explicit in forming a tripartite link between stimulus, receptors, and sensation. First, they predicted the existence of "four types of smell nerves which are stimulated to differing degrees by the various chemical excitants" (325). Second, they related their classification scheme directly to the physicochemical structure of the odorant, in that, for example, the caprylic quality fell off as carbon chain length increased for aldehydes, as did the acid quality, whereas the burnt quality and the fragrant quality also exhibited changes (see fig. 2.2). The eminent experimental psychologist Edwin Boring (1928) was quick to praise this scheme, especially for its reliability—something notably lacking in Henning's scheme—and for its heroism in attempting to introduce order to the vast array of odorous stimuli. Although Crocker and Henderson's (1927) classification system survived until after the Second World War (Crocker 1947), it did nothing to stop either the develop-

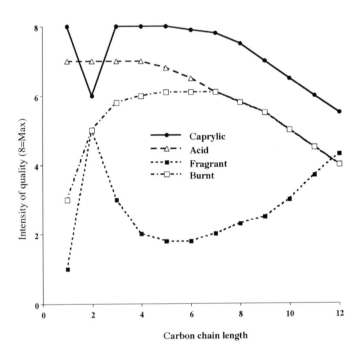

Fig. 2.2. Changes in the four odor qualities of Crocker and Henderson's (1927) classification scheme for aldehydes as carbon chain length increases (data adapted from Crocker and Henderson 1927).

ment of other competing "primary" classification schemes nor, more importantly, the emergence of multiple-descriptor systems that essentially rely on
the similarity method they pioneered (more below). This shift to multiple-
descriptor systems reflects the practical failure of Crocker and Henderson's
(1927) approach, in that it did not serve as a useful tool in communicating
what something smelled like to another person. Needless to say, their prediction of four basic types of receptor does not gel with contemporary findings either.

The classification scheme of Amoore (1952) has amassed the most detailed
body of evidence and is the last true descendant of Henning's approach. Just
as Henning (1924) and Crocker and Henderson (1927) developed pioneering
tools and approaches that outlasted their specific theories, so did Amoore.
First, he derived seven primary olfactory sensations—ethereal, camphoraceous, musky, floral, minty, pungent, and putrid—based on analysis of terms
used in chemistry and perfumery to describe new compounds. Second, he
expended considerable effort exploring specific anosmias (Amoore 1975). In
this case a specific anosmia does not reflect a total inability to smell a particular substance, although this can occur; instead, it indicates a far higher
threshold for detecting the substance than demonstrated in nonanosmic participants. At least initially it looked as if the specific anosmias and the seven
primary odor sensations would nicely tie together to suggest seven specific receptor types. Receptor types were presumed to be based on molecular topography, unlike the earlier two models, which were based on specific structural
features. Amoore's approach is most effectively illustrated by camphoraceous
odors. Most of the molecules that induce a camphorlike smell have a similar
topography, but as is so often the case with any sort of scheme that relies on
one particular characteristic, exceptions can be found (see Rossiter 1996).
Like the other schemes mentioned above, this one too has fallen out with the
available data. At last count there was something approaching 70 specific
anosmias (Amoore 1975). In addition, the same caveats applied to Crocker
and Henderson's (1927) scheme apply with equal force to Amoore's.

The general tendency since the 1970s has been to studiously avoid "primary odors" and, indeed, models in general that attempt to link sensation
with physicochemical (or other) stimulus characteristics. There are several
reasons for this. First, it is in part an acknowledgment that they have done little to further our understanding of olfactory *perception* (note here the stress
on perception). Second, more recent and advanced statistical analysis of
multiple-odor descriptor schemes reveal no underlying structure. That is, the
system is broad and flat, rather than clustered into a few discrete "primary"

sets of qualities (Chastrette, Elmouaffek, and Sauvegrain, 1988). This type of finding is in direct contradiction to any model proposing relatively few odor descriptors or primary sensations (Chastrette 2002). Third, even among experts, there is considerable disagreement about the definition of widely used sets of basic descriptors, for example, in Brud's (1986) nine-category system, although 25% of the perfumers surveyed identified 1-decanol or lauryl alcohol as the archetypal "fatty" smell, the remaining 75% identified 55 different odorants as being the odor archetype. This level of disagreement, which accompanied the other descriptors too, points to a lack of consensus that is striking when compared with judgments of color quality, for example. This suggests that any search for a "basic" set of descriptors is likely to be in vain. Fourth, the creation of sets of "primary qualities" has been of little benefit to the perfume, flavor, or sensory evaluation sectors, who have largely developed their own schemes of classification. These include several different methods of profiling odors. The most common approach is to present the participant with a large number of verbal labels, mainly composed of "odor objects" or descriptors (e.g., smells like . . .). Participants are then asked to evaluate the similarity of the target odor to all the descriptors in the set. Perhaps the most widely known of the several schemes available are those of Harper et al. (1968) with 44 descriptors and Dravnieks et al. (1978) with 146 descriptors and specialist schemes such as Arctander's for perfume (1969) with 88 classes and Noble et al.'s (1987) wine wheel with 90 or so descriptors.

Specific anosmias, selection of primary odors, and searches for systematic relationships between structure and sensation, in general, have been unsuccessful at explaining how we perceive odors. Another experimental psychology approach, which we have not discussed so far but which has the same goal in mind, is specific adaptation (Amoore 1975). The logic of this approach is straightforward. If a particular odorant binds to a particular receptor, then repeatedly exposing that receptor to this odor should selectively fatigue that receptor (or set of receptors). This raises two key questions: are structurally similar odors also adapted out (cross-adaptation) and are structurally different but similar smelling odors affected too? This approach potentially offers a further route to study the relationship between stimulus, receptor, and sensation.

The evidence amassed from several such investigations has been confusing to say the least. Although there are clearly examples in which structurally similar odors cross-adapt (Cain and Polak 1992), there are also examples in which they do not (Todrank, Wysocki, and Beauchamp 1991). Likewise, there are examples of chemicals that have similar structural features and cross-adapt but do not smell alike (Pierce et al. 1995) and examples of odors that do not smell alike but have dissimilar structures and do cross-adapt (Koster 1971). Finally, although self-adaptation will often eliminate, albeit briefly, the ability to smell the target odorant, cross-adaptation has at best a relatively small effect—surprising perhaps if the logic of this technique is pursued to its conclusion—single-receptor fatigue. Not surprisingly, this method has been unsuccessful at annunciating relationships between either physicochemical properties and receptors or between physicochemical properties and sensation; however, more recent work detailed later in this book indicates that adaptation actually provides fairly compelling evidence *against* structure-quality models of olfactory perception.

Although it is of some importance to understand what characteristics of the stimulus yield particular sensations, this is likely to tell us very little about olfactory *perception*. At its most basic, the search for consistent relationships between stimulus and sensation is predicated on the idea that the stimulus produces a set response by activating a particular receptor(s), the response being a sensation that is, all other things being equal (e.g., anosmia, adaptation), *the same* in all participants at all times. However, this simple and apparently useful starting point is undermined when consideration is given to what the olfactory system actually needs to accomplish to "smell." The most difficult problem is how it identifies a biologically significant odor from the array of other odors present at any one particular time. A stimulus-response system is of little value in this respect, because all the system registers is what is presented to it at the receptor level; no attempt can be made with such a system to select a particular pattern of stimulation over that of any other. A related problem also arises. Biologically significant odors are typically not single pure chemicals; rather, they are complex mixtures composed of tens or hundreds of volatile substances (Maarse 1991). Thus, the problem facing the system is even more complex than one first imagines, because the system has to select not just one biologically relevant stimulus but a pattern of stimuli that may themselves change over time and place and that co-occur against a constantly shifting background of stimulation. As we detail below, the implications arising from these two points are fatal to any theory of olfaction that relies solely on a stimulus-response mechanism.

In contrast to the somewhat linear history of psychological thought on olfactory perception outlined above, thinking in sensory physiology has had a more convoluted history—at times closely following the psychological perspective and at other times quite distinct. Although there are important exceptions, recent sensory physiological work clearly seems driven by physicochemical, stimulus-response relationships, whereas psychological work is moving away from this view (see below). The discovery of the large gene family for olfactory G-protein-coupled receptors, and the increasing variety of tools for imaging the odor-evoked activation of individual glomeruli in the olfactory bulb, which receive input from olfactory receptor neurons, has led to the repeated claim in the sensory physiology literature that perceptual outcome (e.g., similarity between odors) can be predicted solely from identification of receptor input and its resulting activation of the olfactory bulb—clearly a stimulus-response view.

This has not always been the sensory physiologist's account of olfaction. In 1942, Adrian concluded, "We recognize a sight not because particular receptors are stimulated but because a particular pattern of activity is aroused, and it is reasonable to conclude that we recognize a smell in the same way. . . . In this way an endless variety of smells might be distinguished because the process would be comparable not to the discrimination of colours but to that of visual patterns" (Adrian 1942, 472). Clearly, receptor transduction and differential sensitivities of olfactory receptors initiate and place constraints on the process, but Adrian seems to argue that olfactory perception involves processes far beyond this initial step. Similar arguments (and data) against the stimulus-response model of olfactory perception have been advanced by Freeman and his co-workers (Freeman 2000). However, a quick overview of olfactory sensory physiology work in the past few years clearly favors an analytical, stimulus-response approach to odor perception.

Problems with This Approach for Object Recognition Tasks

Stimulus-based theories are predicated on the fact that systematic relationships will hold between stimulus and response and that, if these relationships are uncovered, it should be possible to predict what a novel chemical will smell like. There are at least two ways to examine this proposition, from the approach of perceptual ecology and from that of experimental psychology. The world of smell does not occur, in general, as discrete pure chemical stimuli. Biologically important odors, such as those from sexual partners, predators, siblings, offspring, and food, are complex mixtures of chemicals. The primary task of the olfactory system is to recognize these combinations against a background of competing olfactory stimulation. Thus, the olfactory system has to select a meaningful combination of chemicals—an odor object—from the many other possible co-occurring stimuli. A model of olfaction based solely on stimulus-response relationships cannot do this, because it has no capacity to learn the combinations of chemicals that make up each odor object. Of course this objection to stimulus-response models is not in itself fatal. The obvious solution is to tack on a "brain" to extract relevant patterns derived from the receptor array. However, even this solution will not suffice, because it fails to account for the effect that an object recognition system has on all aspects of olfactory information processing. The aim of what follows is to provide a brief overview of why even the "add-on" view is flawed and to thus illustrate that stimulus-response models of whatever form can have no place in a causal explanation of olfactory perception.

Stimulus-response models are analytic, as they presume a particular and discrete response to each component of the stimulus (e.g., one particular functional group evokes one particular sensation, two particular functional groups evoke two particular sensations, and so on). This information may or may not be used, but it is *potentially* available. The first effect that an object recognition system has is to shut off from conscious introspection and behavior information gained from the binding of chemicals to receptors. This is because objects may be composed of many individual chemical components, but because the object is the level at which processing occurs and the one at which biologically significant events occur also, information is only available at this level. The effects of this can be seen in experiments in both humans and animals that have examined whether chemical components can be identified in mixtures of increasing complexity (Laing and Francis 1989; Staubli, Fraser, et al. 1987). In most of these experiments, when the mixture contains more than three or four pure chemicals, the animal or person can not detect all of the original components at better-than-chance level. Of course this could be put down to chemical interactions occurring at the receptor level, yet this observation also holds true, most revealingly, when the "components" are in fact themselves odor objects composed of tens or hundreds of pure chemicals (e.g., chocolate and coffee; Livermore and Laing 1998b). Clearly, this limit on detecting components reflects access limitations to information arising directly from the stimulus-receptor array.

Stimulus-response models presume that relationships between structural features and sensation (or behavior) are rigid, that is, one functional group will evoke the same response in different individuals at different times (setting aside adaptation and specific anosmia). However, object recognition models do not behave like this. First, they rely on learning combinations of chemical stimuli—odor objects. Second, they are capable of redintegration, that is, a component of a complex mixture can evoke the whole and so they degrade "gracefully" and can detect slightly varying or partial scents. The heart of this is that the relationship between stimulus and response can demonstrably break down. This has been observed experimentally in human volunteers. Presenting participants with odor mixtures composed of two elements results in those two elements coming to share odor qualities. For example, a combination of mushroom and lemon odors results in participants later rating the mushroom odor as smelling rather lemony and the lemon odor as smelling rather mushroomy (Stevenson 2001c). This effect has been documented for many different types of odor mixture (Stevenson 2001a; Stevenson, Case, and Boakes 2003) and has been noted by flavor chemists, who, for example, observed that the characteristic bitter almond smell of hydrogen cyanide may arise through its frequent co-occurrence with benzaldehyde (Rossiter 1996). Of course it is possible to dismiss such findings on the basis that they represent nothing more than self-report data, with all their attendant short comings, but the observation that smelling mixtures selectively reduces the discriminability of its components (Stevenson 2001d; Case, Stevenson, and Dempsey 2004) argues against this. Moreover, the finding that reductions in discriminability, alterations in odor quality, and changes in judgments of odor similarity all correlate lawfully with each other provides validity for these self-report data (Case, Stevenson, and Dempsey 2004; Stevenson 2001d; Stevenson, Case, and Boakes, forthcoming). The conclusion here is that encoding odor combinations—objects—results in breakdowns in the relationship between stimulus and response that cannot be predicted from knowledge about the physicochemical or other properties of the stimulus. Yet again, the information available to consciousness does not strictly reflect what is actually being smelled but instead what is actually useful to know.

Not only can breakdowns in the relationship between stimulus and response be seen under these conditions, they can also be seen when a novel odorant is first encountered. If stimulus-response relationships dictate what we smell, then they should evoke the same response when the odorant is first sniffed as when it has been sniffed many times—that is, the same receptors should be stimulated with the same response on each occasion. However, object recognition models rely on the acquisition of patterns of stimulation—odor objects—consequently, responses to odors should change over the first few times that a novel odor is encountered. Exactly such observations have been made, in that novel odors are redolent of many more odor qualities than familiar odors (Stevenson and Dempsey, in preparation) and novel odors are perceived less sharply (Hudson and Distel 2002) and found to be harder to tell apart (Rabin 1988). Similarly, the developmental literature also suggests that learning odor objects affects odor perception, such that younger children are poorer at discriminating odors than older children and adults (Cain et al. 1995; Stevenson, Sundqvist, and Mehmut, submitted). This does not appear to result from maturational effects on the receptor system itself, as otherwise it is difficult to explain how exposure can enhance discriminability to an adultlike level in children (or even neonates) for one odor, while leaving others still poorly discriminated. Yet again, such observations are incompatible with a stimulus-response model.

Stimulus-response models have to presume that, even if the tacked on recognition process were damaged, rudimentary olfactory perception would still take place; that is, information from receptors sensitive to different stimulus features would be available, allowing for some discriminative capacity between differently smelling odors. This is analogous to the observation above, that stimulus-response models should function effectively even with novel odors, yet the empirical data suggest that this is not so. Olfactory neuropsychology provides some of the most compelling evidence that stimulus-response models with an add-on brain can not account for the devastating effect resulting from the selective loss of the memory store of odor objects. Evidence from single case studies of Korsakoff syndrome, Alzheimer disease, and normal ageing all point to the same conclusion, that under conditions where the olfactory object store is lost or damaged, the ability to perceive odor quality is itself similarly lost or damaged (Mair et al. 1980; Eichenbaum, Morton, Potter, and Corkin 1983). In more straightforward terms, these participants can know they are smelling something, they can know that one thing smells more or less strongly than another, but they lack the ability to discriminate between different odorants of the same intensity and relatedly report the absence of odor quality—a rose smells as much like cheese as it does of petrol.

The implication of these neuropsychological findings could not be more stark. Not only do they suggest the importance of an object recognition system, but they also point to its fundamental role in olfactory perception—that conscious perception of smell is based on redolence, the degree to which a pattern of activity from the receptor array resembles different stored patterns of activity encoded in odor memory; no match or no memory equals no perception. That we find relationships between stimulus and discriminative ability and odor quality is not surprising, because these will often have similar patterns of receptor activity. This observation is fundamentally misleading, however, as of course correlations can be, because it falsely suggests a causal relationship between the stimulus and its apparent consequences—sensation and behavior. In fact, as the neuropsychological and learning data both show, this relationship can not be causal; rather, olfactory experience is primarily defined by experience, not by the current content of the receptor array.

In sum, two classes of argument suggest an object recognition system, rather than a stimulus-response model, which has historically been presumed to underpin olfactory perception. First, the ecology of the perceptual system requires it for identifying changeable and novel biologically significant odors composed of multiple components against a changing and complex olfactory background. Second, stimulus-response models, even those that might acknowledge the need for a recognition system, are incompatible with behavioral data and incompatible with the majority of olfactory function. In closing this section it is worth reflecting on the fate of similar models in other areas of perception. Bregman (1990) has argued that in auditory scene analysis, a focus on the psychophysical aspects of audition has contributed little to solving the fundamental problem facing the auditory system, how it comes to extract meaningful patterns of information against a noisy background (see Ullman 1996, for a similar comment with respect to visual object recognition). We would suggest that a similar emphasis on this approach in olfaction—exemplified in our brief historical review—has also diverted attention away from the real problem: how does the olfactory system extract an odor object from a complex olfactory scene?

How Other Systems Perform Object Recognition

The philosophy, psychology, and neurobiology of object perception has a rich literature that extends many hundreds of years—primarily focusing on visual object perception, though including other sensory modalities and multimodal interactions. George Berkeley, for example, in "A Treatise Concerning the Principles of Human Knowledge" originally published in 1710, stated:

By sight I have the idea of light and colours with their several degrees and variations. By touch I perceive, for example, hard and soft, heat and cold, motion and resistance, and all of these more or less either as to quantity or degree. Smelling furnishes us with odours; the palate with tastes, and hearing coveys sound to the mind in all their variety of tone and composition. And as several of these are observed to accompany each other, they come to be marked by one name, and so to be reputed as one thing. Thus, for example, a certain colour, taste, smell, and figure and consistence *having been observed to go together, are accounted one distinct thing*, signified by the name apple. (emphasis added, Berkeley and Dancy 1998, 103)

Given the rich literature on object perception, our explanation and understanding of memory-based odor object perception need not be created *de novo*. Basic tenets, principles, and caveats can be extracted from what we know about object perception in other sensory systems and then *carefully* applied to the unique issues presented by olfaction. The discussion that follows is not intended to imply that object perception in olfaction is identical with object perception in vision (or other sensory systems). It is intended to provide a conceptual framework for understanding how perceptual objects can be created by the central nervous system from a sensory world that is not inherently nor necessarily composed of isolated objects.

The cognitive benefits of dividing the sensory world into perceptual objects are numerous. Perception at the level of objects rather than in piecemeal fashion enhances or contributes to the speed and accuracy of stimulus discrimination (perhaps most notable in the special case of face recognition), figure-ground segmentation, recognition of degraded or obscured stimuli, view-invariant recognition, and perceptual constancy. A downside of perception functioning at the object level is a loss of feature analysis, again perhaps most notable in face recognition.

Several factors influence or facilitate object perception and the formation of perceptual objects. These factors appear to be consistent across sensory modalities where they have been examined. As recognized even by Berkeley, one major factor influencing object formation is temporal coherence. In vision this generally takes the form of coincident movement (Spelke 1990). Multiple features moving in the same direction at the same speed are more likely to be perceptually grouped into a single, coherent object. Object formation in the olfactory system presumably does not involve an external spatial component, but associating temporally co-occurring features is a forte of the nervous system, as recognized decades ago by Donald Hebb (Hebb 1949). Hebb's theoretical cell assemblies and modifiable synapses allowed the nervous system to learn that temporally co-occurring sensory inputs may be intrinsically associated in the form of a single object. This association, mediated by changes in synaptic strength, permitted a partial input experienced later to evoke a complete sensory percept along with any other associated memories. These theoretical processes are now experimentally embodied in cortical cell assemblies described in several central systems and long-term synaptic plasticity, such as long-term potentiation and depression.

The following sections are an overview of what is known about object perception and object coding across several sensory modalities. They are not meant as in-depth reviews of what are very active fields of research; instead, they attempt to extract general principles about how sensory systems encode perceptual objects. Arguments have been made for object perception in vision, audition, somatosensation, and more specialized subsystems such as echolocation. The final section presents our view of odor object formation and its role in olfactory perception.

Vision

Although the processing of visual information begins with a two-dimensional sheet of receptors and, as noted below, there is no necessarily inherent separation between which aspects of the continuous visual pattern should be grouped into distinct objects, our visual system, like that of most other visually guided animals, divides the visual world into perceptual objects. Elizabeth Spelke, a pioneer in visual object perception and its ontogeny in humans, succinctly described the problem facing the visual system and her interest in that problem:

> We have focused on object perception in cluttered, changing arrays . . . [C]luttered arrays are the norm in ordinary environments: Objects rarely stand against a homogeneous medium, separated from one another and continuously, fully in view. More commonly, objects sit upon one another and beside other objects, they are partly hidden by objects closer to the viewer, and they enter and leave the visual field sporadically as the viewer, or some other object, moves. No mechanism for segmenting the surface layout into objects could operate effectively if it could not determine the boundaries between objects that are adjacent, the complete shapes of objects that are partly occluded, and the persisting identity of objects that move out of sight. (Spelke 1990, 30)

There are at least two distinct mechanisms of visual object perception in the animal kingdom. First, in many animals, identification and recognition of visual objects occurs through a process akin to template matching (Sewards and Sewards 2002). An internal, generally innate neural template exists such that, given an appropriate stimulus (e.g., a "fly" or fly-shaped object or a "worm" or worm-shaped object), an appropriate response is evoked. This can be demonstrated in toad visual systems, for example, where a projected rectangle of the appropriate relative dimensions (longer in the horizontal axis than in the vertical axis) and moving in the appropriate direction (parallel to the long axis) evokes capturing and consuming reflexive behaviors. Visual stimuli differing from this general pattern do not evoke such behavior. The perceptual object "worm" for the toad is thus composed of an irreducible combination of features that match some internal, innate template. This is a hard-wired, reflexive system driven by a specific spatial pattern of input. The object cannot be analyzed by the toad into its constituent features—it simply matches the template or does not.

The concept of "sign stimuli" by the Nobel laureate Nikko Tinbergen is a similar example. Sign stimuli are highly constrained, innately recognized visual stimuli that evoke (or increase the probability of) specific behavioral responses. A classic example is the high-contrast, colorful pattern of the herring gull beak. Adult herring gulls have a bright beak with a red spot near the end. Newly hatched gulls instinctively peck at stimuli with these characteristics in the expectation that the parent will regurgitate a meal. Tinbergen identified the specific visual features necessary to evoke chick pecking behavior, essentially identifying the features that fit an innate template representation of a herring gull parental beak in the visual system of the gull chick (Tinbergen and Perdeck 1950). This is a hardwired system entirely driven by the physical characteristics of the stimulus.

The second mechanism of visual object perception involves and requires experience. It is, in a sense, a memory-based process in which experience, often during early development, trains the visual system to extract visual objects from within a scene. There may be innate rules and/or biases in object perception and internal representation, but experience is necessary to learn that a particular visual pattern on the two-dimensional retina represents a three-dimensional object. Experience-dependent object perception enhances the ability to identify specific visual patterns as individual objects, discriminate between similar visual objects, and recognize and perceptually complete objects even if they are partially obscured.

Visual face recognition is a classic example of this process. Humans perceive visual information from faces configurally, and recognize familiar faces nearly instantly as a whole. Configural visual processing involves combining the information regarding features (e.g., eyes, nose, and mouth) and the spatial relationships between them into a single, potentially indivisible percept. Thus, the process does not appear to involve sequential analysis of individual

Fig. 2.3. Face recognition can be a highly configural process. These two faces should appear relatively similar. However, by turning the book upside down to view the faces upright, dramatic differences become obvious. This is because viewing inverted faces impairs feature analysis and configural processing becomes dominant (Murray 2004). We argue that odor object perception is an extreme example of reduced analytical and enhanced configural processing.

facial features; instead, it stems from perception of the pattern as a synthetic whole—a gestalt (Tanaka and Farah 2003). The reliance on configural- versus feature-based analytical processing of faces can be manipulated by inverting the faces. At first glance, the two inverted faces shown in figure 2.3 appear very similar. However, flipping the page upside down enhances our access to feature-based analytical information and the two faces become highly distinct. Thus, in this case, we can shift from configural (all the parts are in the right place and the face as a whole appears normal) to analytical perception. Perception of nonface visual objects (cars, animals, chairs) falls somewhere along the continuum between configural and analytical, in part, depending on the familiarity of the objects and the expertise of the viewer for that class of objects (Tanaka and Farah 2003).

A major difference between configural perception of right-side-up faces and feature-bound perception of inverted faces is a reduced attention to and recognition of individual features during configural processing. Thus, subtle changes in individual features of a familiar face can go unnoticed in a configural percept, and the more familiar the face, the stronger the effect.

Special brain regions may be involved in configural face perception or configural perception of objects for which the viewer is an expert. The fusiform face area (FFA), for example, is differentially activated when humans are viewing right-side-up faces as compared with perception of other object classes. Perception of most other objects seems to be mediated by activity within the inferotemporal cortex. Single neurons in the visual inferotemporal cortex express receptive fields for complete visual objects and can alter those receptive fields based on experience. The data used by the inferotemporal circuits comes from the highly analytical periphery (retina and visual thalamus) and hierarchical processing in early cortical stages. Apparently, there is an organization of object encoding in the inferotemporal cortex wherein cells responding to similar objects are located near to each other (e.g., regions of face-selective cells and regions of non-face object-selective cells). This spatial organization and experience-dependent plasticity of intracortical association connections may contribute to the development of "view-invariant" neurons, which can respond to a three-dimensional object regardless of the angle of view (Sejnowski 1996; Tovee, Rolls, and Ramachandran 1996).

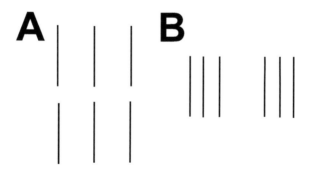

Fig. 2.4. Two examples of stimuli used in visual perceptual learning. (A) In vernier acuity testing, subjects are required to determine whether the two vertical lines are precisely in line. (B) In three-line bisection tasks, subjects are required to determine whether the center line falls precisely in the center between the two outer lines. Performance accuracy (perceptual acuity) in both of these tasks can be improved dramatically with practice.

As noted above, visual perception is experience dependent, as is the underlying neural processing. Experience can influence both visual feature coding and configural coding of objects. Visual acuity and discrimination of fine visual details can be enhanced by training and experience, a process called perceptual learning. Vernier acuity, for example, the ability to discriminate whether two vertical lines are precisely in line or are slightly horizontally displaced, can be dramatically improved with training. The enhanced acuity for vertical lines does not generalize to different orientations; thus, if trained on vertical lines, there would be no improvement in discrimination performance for horizontal lines. Furthermore, if training is limited to a specific region of the visual field or to one eye, the training does not transfer to the other regions or the other eye. These limitations strongly suggest that the neural substrates for perceptual learning of visual feature acuity occur relatively early in the visual stream, prior to, for example, the inferotemporal cortex where receptive fields are very large and bilateral. In fact, recent functional magnetic resonance imaging (fMRI) data in humans (Schwartz, Maquet, and Frith 2002) and single-unit recordings in monkeys (Crist, Li, and Gilbert 2001) implicate changes within the primary visual cortex as underlying perceptual learning of feature acuity.

Experience has two effects on configural visual object encoding. One effect is at the level of feature encoding. Encoding of the component features of visual objects by inferotemporal cortical neurons becomes more selective with familiarity. Thus, neurons that may have originally responded to a variety of similar object features become more focused in their response range when viewing familiar objects. The second effect of experience on visual object encoding by single inferotemporal neurons is that responses to the complete object become more selective. Thus, neurons that may have originally responded to several similar visual objects and/or their component features become more narrow in their responding, giving their most robust response to the entire unique familiar object and responding much less to similar objects or object components (Baker, Behrmann, and Olson 2002). Therefore, individual neurons become tuned to a particular familiar object or object class through experience.

A final example related to visual object perception is the ability to perceive an object as being constant despite changes in orientation due to movement of the object and/or the viewer. There are a variety of theoretical mechanisms of object invariance including work by Marr and more recently by Biederman and colleagues (Marr 1982; Biederman and Gerhardstein 1993). A basic tenet of most of these theories is an experience-dependent change in intracortical association connections that allow convergent buildup of multiple representations of the same object from multiple viewpoints on individual target cells or local ensembles. Thus, on initial viewing, a neuron in the inferotemporal cortex may respond strongly to one view of an object, but less so to views of the same object from different angles. With exposure to the object from different views, in general, occurring as the object or viewer moves around an axis, inferotemporal neurons can develop view-invariant receptive fields for the object, responding consistently regardless of the viewing angle (Tovee, Rolls, and Ramachandran 1996). Similarly, behavioral perception and recognition of objects viewed from different angles is enhanced as experience viewing those objects increases.

Together, these experience-dependent processes enhance (1) discriminability of similar visual objects, (2) configural perception, which allows rapid recognition of familiar whole objects but may interfere with local feature perception, (3) view-invariant perception, which facilitates object constancy and recognition of distorted or obscured images, and (4) recognition of the visual object from background. In subsequent chapters we argue that substantial conceptual, if not mechanistic, similarities exist between visual object perception and olfactory perception.

Note that object perception can be guided and influenced by top-down processing and expectation. Mechanisms of the effects of expectation on visual object perception and recognition are not fully understood but clearly involve memory circuits. Top-down processing occurs at nearly all levels of the visual pathway, with, for example, more inputs to the visual thalamus coming from the visual cortex than from the retina. A similar role of top-down processing and descending activity is proposed for the olfactory system, as discussed elsewhere in this volume.

Somatosensory System

The other sensory system most commonly used by humans to identify and discriminate objects is the somatosensory system. Using one's hands to explore and identify objects is known as haptic object perception. There are strong multimodal interactions between visual and haptic object perception, wherein, for example, familiarity with an object through haptic exploration can enhance visual recognition of that object and visa versa (e.g., Norman et al. 2004). In fact, recent fMRI work suggests that the visual object perception areas in the inferotemporal cortex are also activated by haptic exploration.

One possible explanation is that exploring an object by hand may induce a visual image of that object. However, the same region is activated by haptic exploration in congenitally or early-onset blind subjects, suggesting instead a true multimodal role for the inferotemporal cortex in object perception (Pietrini et al. 2004).

In vision, several basic principles guide object perception, including (1) features that move coherently are more likely to be perceived as belonging to a single object, (2) features with similar visual texture and/or color are more likely to be perceived as belonging to a single object, and (3) features that appear to fall along a continuous smooth contour are more likely to be perceived as belonging to a single object. Similar rules appear to guide haptic object perception (Spelke 1990). In vision, however, the spatial scanning of a scene allows spatiotemporal coherences to be extracted and applied to the object perception task, but in haptic object perception moving the hands along the surface and edges of the stimulus object extracts the required spatiotemporal information. This movement (often guided by vision as noted above) allows central circuits to learn a configural representation of the object. As in vision, configural haptic object perception enhances recognition of familiar objects, even if only partially accessible, and enhances discrimination between similar objects (Spelke 1990).

Object perception has been examined in other sensory systems including the auditory system. Echolocation, for example, used by bats and marine mammals can provide intricate detail of object features, which can then be used to identify objects (Harley, Roitblat, and Nachtigall 1996). Object perception in echolocating bats can be very similar to visual object perception where an "object" may include an insect or flower, whereas in auditory situations that humans are more familiar with, auditory "objects" may include the voice of a particular speaker against some background noise or against other speaker objects. Being able to segregate sounds according to source allows following one speaker's voice and verbal message, although multiple sounds are impinging on the cochlea simultaneously (the cocktail party effect). Similar basic rules of spatiotemporal coherence and similarity in features appear to guide object formation in audition as in the other systems described. As in other sensory systems, auditory object perception enhances discriminability of similar objects and enhances separation of objects from the background.

The final point about auditory object perception (where an auditory object represents the unitary source of the sound) is that it can be strongly in-

fluenced by vision. Thus, as in haptic object perception, multimodal interactions are important for complete and accurate auditory object perception.

GENERAL PRINCIPLES

This very superficial review of object perception across sensory systems serves
to emphasize several points of relevance to our discussion of olfactory perception. First, object perception is a basic phenomenon of most, if not all,
sensory systems. Despite highly analytical peripheral mechanisms and cellular receptive fields (e.g., for spots of light at specific spatial locations in vision,
for mechanical deformations or vibrations at specific spatial locations in somatosensation, and for frequency tuning in audition), higher cortical sites in
all of these systems perform configural or synthetic processing that helps create perceptual objects. Object perception may be driven by either innate,
hard-wired systems evolved to recognize highly specific stimuli (e.g., sign
stimuli), or may be driven by experience- or memory-based object perception. Memory-based object perception may include innate biases for perceiving some kinds of objects over others (e.g., potentially faces in humans,
though this is debatable), but in general, memory-based perception can significantly enhance the breadth and flexibility of perceptual abilities.

Second, object perception is frequently, perhaps usually, multimodal.
Strong interactions occur between vision and both haptic and auditory object perception. Visual recognition of an object, for example, can influence
haptic perception of that object. Similarly, the ventriloquist effect (locating
the sound source at the moving mouth of ventriloquist's dummy) is a prime
example of how vision can shape our perception of auditory sources and objects.

Finally, given the ubiquity of memory-based, synthetic object perception,
one may be led to believe that there is an adaptive advantage or benefit to
such processing. In fact, several advantages of synthetic/configural object
perception over local feature perception are apparent from the examples provided above. First, synthetic object perception allows very rapid identification and recognition of objects and discrimination of those objects from similar objects, without requiring a feature-by-feature analysis. Indeed, synthetic
object perception can actually suppress perception of component features.
Second, synthetic object perception enhances identification of familiar objects from background stimuli. As described by Spelke (1990), given the in-

credible complexity of most sensory scenes (in all sensory modalities) and the reliance on relatively simple receptor sheets to transduce and begin analyzing those scenes, the ability of higher order central circuits to group and organize the scene into distinct objects and background is fundamental to effective perception. Finally, synthetic object perception can facilitate perceptual constancy. Viewing (or feeling or hearing) a familiar object (it must be familiar) from different vantage points can create remarkably different spatio-temporal inputs to our sensory systems, yet we are still able to perceive that object as a constant, immutable whole. The soda bottle next to my laptop as I type looks like a soda bottle when I stand up or sit down or even when the monitor partially obscures it, despite the extreme changes these different views create in the image striking my retina. Synthetic object perception and its central coding allow this robust constancy in the face of image distortion and occlusion.

OLFACTION AS A MEMORY-BASED, OBJECT RECOGNITION SYSTEM

We hypothesize that most animals (vertebrate and invertebrate) have two modes of olfactory perception. One mode consists of a physicochemically driven, hard-wired (labeled line?) process under strong evolutionary control and most generally used for perception of generationally stable, adaptive stimuli such as pheromones, predator or host odors, and perhaps some food odors (such as amino acid detection in fish). Detection of an appropriate odor ligand, regardless of the presence of other stimuli, increases the probability of specific behavioral responses. An example is the suckling pheromone described in neonatal rabbits (Schaal et al. 2003). A specific volatile odorant from rabbit doe nipples, 2-methylbut-2-enal, produces behavioral activation and suckling behaviors in newborn rabbits but not in rodent neonates. This compound is one of many dozens of volatiles released by the nipple, yet "nipple odor" as a whole is not required to express the behavior—exposure to 2-methylbut-2-enal in isolation is sufficient.

In vertebrates, physicochemically driven perception has often been seen as the role of the vomeronasal and accessory olfactory system; however, increasing evidence suggests an important contribution of the main olfactory system. In the physicochemical mode, knowing what physicochemical stimulus is present and/or what specific central circuits are activated provides strong predictive power for determining the olfactory percept (maternal nipple, predator, etc). Although this mode allows for accurate identification of specific odorants and reliable, rapid responses, it is necessarily limited in its tuning breadth to evolutionarily stable stimuli with predictable meanings.

The second olfactory perceptual mode is a synthetic, memory-based mode that rapidly learns to form perceptual odor objects from variable, novel patterns of input. This mode of processing allows perceptual grouping of the complex, feature-rich input extracted by the highly analytical receptor sheet into relatively discrete odor objects, distinct from other patterns of input and distinct from background odorants. Most odors experienced by animals are complex mixtures of molecules, which themselves are composed of multiple submolecular features. In addition, very different odors can have extensively overlapping features, and, in general, odors are experienced against an odorous background. The task of the memory-based processing mode, then, is to learn which features should be grouped (associated) together to form a perceptual object. Once this associative learning has occurred and perceptual odor objects are formed, discrimination, recognition, and figure-ground separation of those objects from other objects and background are enhanced. Although the extent of overlap or spatial similarity of central maps may place limits or constraints on perceptual grouping and ultimate formation of distinct odor objects, knowledge of the physicochemical features or olfactory bulb maps evoked by those features alone is insufficient to predict the ultimate olfactory percept. The strength of this synthetic processing mode is that novel stimuli of varying complexity can come to acquire a unitary percept (and meaning), vastly extending the tuning range of the olfactory system. It is the synthetic processing mode that provides us with the unitary olfactory percepts (odor objects) of "coffee," "fresh baked bread," and Chanel No. 5, despite the fact that these odors are composed of hundreds of individual components. The similarity of this process and its underlying circuitry to visual face recognition is striking (Haberly 2001).

Experience can shape odor coding in at least two ways. First, encoding of odorant features may improve or change with experience, such that familiar features are more precisely or fully encoded. We believe that olfactory bulb circuits encode and enhance representation of odorant features and thus predict that mitral cell encoding of familiar features will be modified compared with novel features. Second, experience can shape odor coding through synthesis of co-occurring features into odor objects. Simple convergence and coincidence detection of co-occurring features by cortical neurons is not sufficient to account for the complex nature of synthetic odor perception. Rather, we and others hypothesize that piriform cortical circuits learn which odorant features co-occur and, through associative synaptic plasticity, store a representation of that feature combination. Once this representation is stored it is more easily recognized from other, similar patterns of input and is robust in the face of partial degradation. This combination of characteristics, as noted above, makes familiar odor objects more distinct from other stimuli, easier to recognize against a background, and also results in a break from strict reliance on olfactory bulb maps of odorant features for a complete olfactory percept. The anatomy of the piriform cortex and the hypothesized reliance of the learning process on broadly projecting intracortical association fibers make it likely that odor object percepts are ultimately encoded by distributed ensembles of cortical neurons, without a precise organization or map of odor objects. Again, this is similar to representation of visual objects in the inferotemporal cortex.

Synthetic formation of perceptual odor objects may not only be built from multiple molecular and submolecular volatile features but also may include multimodal components and/or be influenced by cross-modal interactions. Although vision and somatosensation, and vision and audition appear to have unique, special cross-modal access in the formation of perceptual objects in those systems, olfaction and gustation (and perhaps trigeminal) information may have a similar special relationship. For example, familiar odor objects can include "sweet" components (Stevenson, Prescott, and Boakes 1995).

Finally, inclusion of memory in olfactory object perception raises the importance of top-down processing in this system. We hypothesize that, as in vision and other sensory systems, expectation and internal behavioral state can influence odor perception. Odor object percepts, therefore, may not only include multiple odorant features and multi-modal components, but also may be guided by past associations, expectations, verbal labels, etc. The impact of descending and centrifugal fibers is apparent throughout the olfactory system, including at the glomerular layer of the olfactory bulb itself.

These characteristics of the memory-based, synthetic olfactory processing mode, most well supported by experimental data to be described in later chapters, accentuate the distinction that must be made between identification of the physicochemical features of an odorant and the ultimate resulting odor perception. With a relatively few exceptions, neither odor physicochemical feature extraction at the receptor sheet, nor spatial maps of those features in the olfactory bulb, nor simple convergence of those features in cortical circuits are sufficient to account for the rich experience that is olfaction.

3

Receptive Mechanisms

Odor perception is the consequence of computations occurring within the primary olfactory structures and interactions within and between a myriad of other nonolfactory systems. As a starting point for understanding how the olfactory system functions, we will examine two things. First, we must have a firm understanding of the olfactory stimulus and how stimulus information is acquired by the olfactory system. For example, understanding acoustic waves has been fundamental to our current views of how the auditory system works (though, as noted above, it is not sufficient in itself for a complete understanding of auditory perception). What are olfactory stimuli, how do they differ from stimuli processed by other sensory systems, how are they influenced or shaped by the environment, and what constitutes an olfactory scene (if olfactory scenes even exist)? Understanding the stimulus and how the stimulus is used by the receiver to direct behavior—the ecology of olfactory perception—is critical to understanding the transform functions that must take place within the olfactory system to allow this process to occur. Second, we must have a firm understanding of olfactory system circuit anatomy. As with an exploration of the nature of the stimulus, exploring olfactory system circuit anatomy can help identify possible mechanistic opportunities and constraints on circuit function and its output.

However, before we begin our overview of odors and olfactory functional anatomy, we will briefly perform a similar analysis of a comparator system —the visual system. We could have chosen other sensory systems, but the visual system is probably the best understood by the most people and thus should be most useful for our purposes. We will not deliver a detailed, nuanced description of the mammalian visual system; instead, we will identify critical functional components that we can then try to find analogies for in olfaction.

The Visual System

At one level, of course, the stimulus driving the visual system is electromagnetic radiation within a specific range of wavelengths and varying in intensity. If this were all the useful information carried by visual stimuli, then a small sheet of photoreceptors, varying in sensitivity to wavelength and intensity might be a necessary and sufficient visual apparatus. In fact, some invertebrates have very simple eyes (ocelli) containing a handful of photoreceptors sensitive to wavelength and intensity and not much else. Add some local or central circuitry to enhance contrast between similar wavelengths and discrimination within the electromagnetic spectrum becomes rather efficient. Most visual systems do not function as light meters, however, providing a direct readout of wavelength and intensity; instead, they provide relative information and are most sensitive to change rather than to stable stimuli of a fixed wavelength and intensity.

Obviously in addition to wavelength and intensity, however, visual stimuli can also include spatial information. Some wavelengths at a certain luminance come from over there, whereas different wavelengths at a different luminance come from over there. Processing information in this spatial dimension adds two important layers of complexity. First, circuitry must exist that allows discrimination along the spatial dimension. This is most commonly handled with a larger receptor sheet (or multiple small sets of receptors as in an invertebrate compound eye) and topographic projections from the receptor sheet to central circuits that maintain, in some form, the spatial organization of the receptor sheet (e.g., retinotopic projection). Lateral inhibition and differing levels of convergence help define and accentuate specific spatial patterns of visual stimulation.

The second issue that may arise when a spatial dimension is added to our visual system is the need to direct attention or processing power differentially over the visual field. This can be done in a hard-wired manner, for example, in the visual fovea, where photoreceptors are densely packed and convergence of photoreceptors to second-order neurons is limited. The retinal periphery, in contrast, has high levels of receptor to retinal output neuron convergence. This produces a relatively enhanced proportion of the visual pathway devoted to processing signals emanating from the fovea compared with more peripheral regions. Another mechanism for directing differential processing power to some regions of visual input over others is a dynamic process involving either movement of the receptor sheet in space (e.g., saccadic eye movements), or virtual movement of attention over the visual input via changes in central circuit function (e.g., a processing "searchlight" [Crick 1984]).

However, even a system capable of wavelength and intensity discrimination, spatial discrimination, and directed attention does not come close to describing how we (and many animals) use our visual systems. In fact, at a conscious level, absolute wavelength and intensity are often lost to us. Our visual systems are used for discriminating objects—cars and chairs and faces and trees (see fig. 3.1). Furthermore, our visual systems are robust object discriminators, in particular, when discriminating familiar objects. Chances are you can recognize the visual object that is your car in the parking lot at noon in full sunlight, at noon under cloudy skies, at sunset and even at dusk; and from the side, the front, and the back; and from 100 meters across the parking lot and from within 10 meters—all despite dramatic variation in wavelength, luminosity, spatial size, and pattern. This is especially true (often mistakenly) if we expect to see our car in that lot. All of these conditions produce very different patterns of activity at the receptor sheet yet are recognized as one and the same object at more central structures. Visual scene analysis requires extensive, experience-dependent processing well beyond the basic wavelength, luminance, and spatial discrimination occurring in the retina.

The mammalian visual system circuitry underlying the characteristics described above contains hierarchical processing stages, parallel processing streams, and extensive feedback pathways. In the retina, subclasses of photoreceptors expressing different photosensitive proteins are differentially sensitive to wavelength and are also differentially sensitive to intensity, with one class of receptors sensitive to very low levels of illumination and the other having higher light thresholds. These different types of photoreceptors differentially communicate with retinal output neurons, which can also be divided into different classes. Slowly adapting parvocellular retinal output neurons receive input primarily from high-threshold, wavelength-discriminating cone photoreceptors, and rapidly adapting magnocellular output neurons receive input primarily from low-threshold, rod photoreceptors. The different physiology of these parvocellular and magnocellular output neurons results in par-

Fig. 3.1. This two-dimensional image is viewed as a series of objects against backgrounds rather than as a single contiguous object. For example, the tissue box in the background is perceived as a three-dimensional box extending to the ground, despite the foreground object blocking our view. This segmentation of the visual world into objects depends on experience, which largely occurs during development. A variety of perceptual rules have been identified that facilitate this segmentation of the world into unique objects, as discussed in the text. Photo courtesy of Tristan Sullivan-Wilson.

allel visual information streams, one largely dealing with spatial detail and color and the other processing spatial movement. These parallel streams are largely segregated throughout the visual pathway, terminating in ventral higher-order visual cortices (parvocellular pathway and the inferotemporal cortex) and medial higher-order visual cortices (magnocellular pathway and medial temporal cortex).

Throughout both of these parallel pathways, hierarchical processing occurs, wherein simple visual features extracted at the retina are synthesized into more complex visual objects through convergence along the pathway. This is the classic work began by Hubel and Wiesel wherein, for example, retinal output neurons and their target cells in the lateral geniculate nucleus of the thalamus have monocular spatial receptive fields for spots of light at specific locations in the visual field, whereas neurons in the primary visual cortex have receptive fields for contrast edges or a bar of light (synthesized from many spots) (see fig. 3.2). Ultimately, in the inferotemporal cortex, single neurons may have receptive fields for complete visual objects or faces and continue to respond even if the object is rotated or changes size. These view-invariant receptive fields of inferotemporal neurons, however, require prior experience of viewing the object at different angles to be expressed (Tovee, Rolls, and Ramachandran 1996).

Lateral and feedback interactions are critical for circuit function both within individual stages of the visual pathway and between higher and lower stages. Lateral inhibition between retinal neurons creates the center-surround nature of their receptive fields and in doing so enhances contrast between both similar spatial and similar wavelength inputs. Lateral excitatory and inhibitory interactions between cortical neurons help reinforce columnar processing of visual features. More long-range intracortical associational connections accentuate temporal synchrony of disparate columnar units, con-

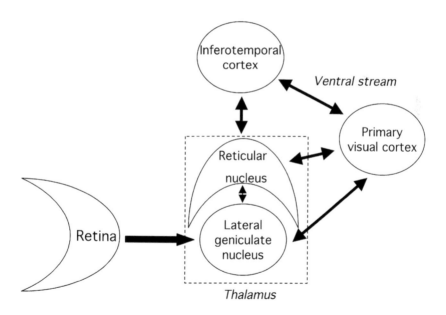

Fig. 3.2. Highly schematized diagram of information flow underlying visual object perception. Information output from each stage becomes more complex in a roughly hierarchical manner as one moves from the periphery to inferotemporal cortex. However, there is extensive feedback at each level allowing for experience, expectancy, and context to modulate activity at earlier stages and ultimately to modulate perception.

tributing to synthesis of features encoded by those columns into visual objects by target neurons in the higher-order cortex. The strength of these synaptic connections is highly experience dependent, expressing both long-term potentiation and depression and thus creating an experience-dependent perception of visual stimulation.

Besides the lateral excitatory and inhibitory connections within specific structures, there are extensive feedback connections between them. For example, the visual lateral geniculate nucleus of the thalamus receives more axonal input from the visual cortex than it does from the retina. Thus, rather than serving as a passive relay of visual information from the retina, the visual thalamus processes retinal input "in the light of" what the visual cortex already knows. Feedback between higher and lower stages occurs throughout the myriad of visual cortical regions (Kastner and Ungerleider 2000).

In addition to (and in part through) helping enhance contrast between objects and suppress extraneous information, these feedback pathways may be the structural basis of the attentional searchlight mentioned above. Through feedback pathways, how incoming information is processed can be influenced by what has already been processed. Our attention for details within a visual scene can be influenced by what we have already gleaned from that visual scene and/or by our expectations regarding the scene.

One of these feedback pathways in mammals deserves special consideration here. The reticular thalamus (thalamic reticular nucleus) is a thin lamina of neurons along the superficial edge of the thalamus and is modality and topographically organized. Thus, for example, a subset of reticular thalamic neurons receive input from the visual cortex and project to the visual lateral geniculate nucleus neurons in a retinotopic manner (Guillery, Feig, and Lozsadi 1998; McAlonan and Brown 2002). The reticular thalamus receives excitatory input from both the sensory thalamus and sensory cortex and projects back to the sensory thalamus. This feedback to the sensory thalamus is inhibitory. Thus, the reticular thalamus is a group of inhibitory neurons, receiving cortical feedback (as well as state-dependent modulatory inputs from basal ganglia and locus coeruleus) and capable of gating/filtering the flow of thalamic information to the cortex. It has been hypothesized (Crick 1984; McAlonan and Brown 2002) that the reticular thalamus plays a role in attentional changes in sensory coding, i.e., serves as a searchlight to enhance processing of important information while suppressing processing of unimportant or unexpected information. In fact, lesions of the reticular thalamus do not impair basic visual behaviors but do disrupt visual priming—an enhanced attention to a specific visual cue based on past experience and expectation (Weese, Phillips, and Brown 1999).

Finally, as mentioned above, attention can also be directed by physical movement of the retina and, thus, differential spatial sampling. Ocular reflexes are mediated by the superior colliculus, which receives input directly from the retina, as well as the visual cortex and the eye fields of the supplementary motor cortex. Among other things, ocular reflexes allow fast movement (saccades) of the eyes to fixate the fovea on a new region of interest and allow smooth eye movements to allow the fovea to maintain fixation on a moving object of interest. Thus, this is a circuit to direct visual attention toward objects or regions of interest through changes in sampling parameters, i.e., physically moving the eyes.

Together, the initial peripheral coding of wavelength, intensity and spatial patterns, the hierarchical and experience-dependent synthesis of visual features into visual objects, and the gating and filtering of information flow through attentional mechanisms results in visual perception. Importantly, a technical description of the spatial patterns of wavelength and luminosity of a stimulus or of the patterns of retinal activation they produce are not sufficient to predict what will be perceived by the viewer.

The Olfactory System

Stimulus and Stimulus Acquisition

As with vision, the stimulus driving olfactory system activity can also be described in fairly simple terms. For terrestrial animals, in general, odorants are relatively small, volatile molecules, varying in size, conformation, and number and location of functional groups. Subtle changes in structure, for example, isomerization, can produce rather dramatic changes in odor percept. A collection of chemoreceptors, each with binding sites selective for different molecular configurations, could be sufficient to account for olfaction with this view of the stimulus. In fact, under strong evolutionary pressure, highly selective chemoreceptors have evolved for generationally stable pheromones.

However, having a single odorant–single receptor system could limit the range of odorants to which the system could respond and severely limit responsiveness to novel compounds. Furthermore, to enhance the information content of chemical signals, yet limit energetic costs of macromolecular synthesis, many behaviorally relevant chemical signals are mixtures of small, simple molecules (described in more detail in chapter 5). Thus, processing of these mixtures requires, or minimally would be facilitated by, olfactory receptors and initial information processing emphasizing simple feature detection followed by synthesis of those features into complete, more complex, perceptual objects. A feature extraction and synthetic processing mode like this would also expand the ability of the system to respond to novel compounds. If the receptor repertoire was large, the diversity of odorants to which the system was potentially responsive could expand enormously. In fact, as we now know, the receptor repertoire of mammals can reach the hundreds, and even invertebrates can express scores of different receptor genes. This results in a palette of distinct odor perceptual objects for humans believed to be in the tens of thousands (at least), far beyond the number of individual types of receptors (see fig. 3.3).

The other hypothesized consequences of synthetic object perception in olfaction, as in vision, are that familiar objects can be perceived or recognized despite partial pattern perturbation or degradation (perceptual constancy), and expectancy and context can shape our perception of odor objects. This can be observed, for example, in a situation where there is some overlap in molecular features of the background odor in a room and some target odor. Imagine a room containing hints of new carpet smell and potted plants. These odors may be detectable upon entering the room but are quickly filtered by central adaptation. Several molecular features of these odors may overlap with coffee or cologne, yet, if these target odors are introduced into this situation, they can be easily identified despite having some features missing because of existing adaptation (assuming they are already familiar) (Kelliher et al. 2003; Goyert et al. 2005). Another example of robust perception in the face of changes in patterned input is the ability to recognize many odors as single stimuli over a wide range of intensities (although the perceptual quality of some odorants changes dramatically with change in concentration; see chapter 4). These examples may be comparable to the visual examples above involving changes in luminosity and object size with changes in distance, vantage, and/or time of day. Finally, as described elsewhere in this volume, even expert wine tasters can be perceptually fooled by tinting white wine red (Morrot, Brochet, and Dubourdieu 2001); that is, how the volatile components in the headspace of a glass of wine are processed by the olfactory system and perceived can be influenced by the expectations of the smeller.

As noted above, the synthesis of perceptual odor objects, unless evolu-

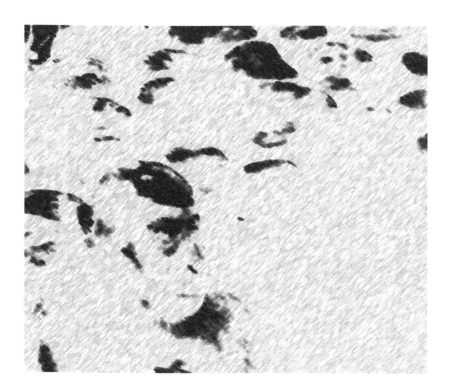

Fig. 3.3. This image is meant to represent the variety of feature information extracted by the olfactory receptor sheet in response to inhaling odorized air. Each black form thus may represent a molecular feature of the odorants present in the inhaled sample. Through a large set of olfactory receptors, the peripheral olfactory system and initial olfactory bulb circuits identify and discriminate between these various features. Olfactory perception, however, requires a configural component wherein these features are synthesized into odor objects. This synthetic processing requires experience. An example of how experience can influence synthetic processing of stimulus features can be seen by examining figure 3.5, then looking at this figure again.

tionarily hard wired, is experience dependent. We hypothesize that the olfactory system learns which features tend to go together in much the same way that visual objects are learned by the visual system by synthesizing visual features that appear or move together. However, object synthesis, both in vision and olfaction can lead to impaired recognition of, or attention to, object features. Thus, subtle changes in visual facial features can be difficult to detect in familiar faces because the face is perceived, and acted upon, as a gestalt—a unique, single object distinct from its components. In olfaction, this disregard of object component features is so extreme that subjects cannot identify all of the components of mixtures containing more than three or so odorants (Laing and Francis 1989). In addition, individual odors can come to acquire perceptual qualities of odors with which they have been associated (Stevenson 2001a). Furthermore, this experience-dependence of odor object synthesis creates an opportunity for odor objects to include non-olfactory components. Thus, for example, familiar odors may include an apparently gustatory sweet component (Stevenson, Prescott, and Boakes 1995).

As a final comparison with vision, odors, unlike visual stimuli, do not intrinsically contain a spatial component, though they may vary in intensity over space as they diffuse from their source. Thus, a strict analogy with visual scene analysis may seem inappropriate. However, an examination of how odors are used by animals (and humans) clearly suggests that something akin to scene analysis can be performed with olfaction. For example, when approaching a table covered with a variety of aromatic, tasty dishes while blindfolded, the scent of some foods will be easily recognized and identifiable (e.g., roast beef), despite the presence of multiple competing odors (broccoli, pasta with parmigiano cheese, and a glass of Cote de Rhone). That is, from this complex olfactory scene of spatially discrete odor sources, some patterns of input can out-compete others and be recognized as distinct wholes. This effect is strongly influenced by relative intensity of the odors (Kay, Lowry, and Jacobs 2003) but also clearly demonstrates that despite the fact that on any given inhalation molecular features from many odor objects are sensed, what is perceived is driven by processes functioning at the level of individual, robust odor objects. The stability of familiar perceptual objects in the face of competing inputs and changes in intensity is a strong adaptive advantage of this processing system.

Thus, from a distance, relative object intensities facilitate identification of odor sources in the scene. However, it is also possible, upon reaching the table (still blindfolded), to spatially sample the area, differentially casting one's nose back and forth across the table, actively sniffing at some points and not at others, and despite the mixing of odorants from different sources that occurs, identify each familiar odor in turn. At present it is unclear if active sniffing and spatial sampling are strictly analogous to saccadic eye or head movements in vision, though sniffing appears to be indicative of attention and certainly allows differential sampling of an olfactory scene in much the same way as eye movements across a visual scene. Neural control of active sniffing (or correlates in invertebrate systems), and the functional consequences of active sniffing on odor coding and perception have been woefully understudied, despite scattered intriguing results (Heinbockel, Christensen, and Hildebrand 1999; Young and Wilson 1999; Sobel et al. 2001) and the obvious potential importance. As the table is olfactorily scanned, molecules from the roast beef will be present during sniffs over the Cote de Rhone, but, we argue, because of (1) the enhanced intensity of Cote de Rhone features while sampling directly over the glass, (2) central adaptation to features of odor objects sampled immediately prior to sampling over the glass (e.g., roast beef), and (3) the robust nature of odor object perception, Cote de Rhone will be identifiable as a distinct object in the olfactory scene. This process is quite different from identifying individual components of a complex mixture that turns on and off as a whole. The temporal component of the type of olfactory scene analysis described here is critical for the scene and object analysis proposed (Jinks et al. 1998). As in vision, dynamic coherence of a collection of features is crucial for perceiving those features as a synthetic object distinct from background. For example, movement of a V-shaped pattern of dots against a static background of dots can lead to the perception of a single arrowhead against a dotted background. In olfaction, we propose that the differential temporal emergence of specific collections of odorant features as we sweep our nose over the broccoli and then the Cote de Rhone assists in perceptual organization of those features into individual familiar odor objects distinct from other objects in the scene.

The point here is that understanding how molecular features are encoded and interact at peripheral stages of olfactory processing is only the tip of the iceberg that represents olfactory perception. We must not lose sight of the fact that, in the same way that wavelength and luminosity are critical for vision but alone would represent a very impoverished sensory modality, the analytical chemistry occurring at the early stages of the olfactory system does not represent the richness of information content and perceptual experience that is olfaction. The question remains, how is olfactory perception accomplished? Below is a brief description of the current state of knowledge of olfactory system circuitry that underlies odor perception (fig. 3.4). An understanding of circuitry often helps define function and identify possibilities and limitations for information processing.

Circuit Anatomy

Many outstanding reviews of olfactory system anatomy exist in the literature. Almost without exception, these reviews follow a common pattern by de-

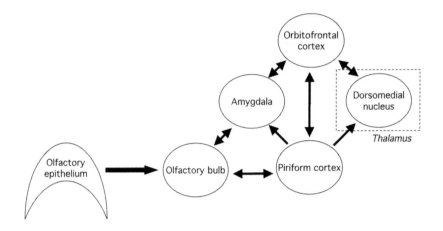

Fig. 3.4. Highly schematized diagram of information flow underlying olfactory perception. The olfactory system includes both a direct projection to the primary sensory cortex without an intervening thalamic relay, as well as a thalamocortical loop. Information flows from the highly analytical olfactory epithelium and glomerular layer of the olfactory bulb to highly synthetic cortical regions such as piriform cortex and orbitofrontal cortex. The olfactory pathway is also highly interconnected with limbic structures such as amygdala and hippocampus. As in vision, there is extensive feedback at each level, allowing for experience, expectancy, and context to modulate activity at earlier stages and ultimately to modulate perception.

scribing the system from the periphery to higher central structures. This organization makes sense because, in general, this is the initial direction of information flow. However, a description organized in this manner also fosters assumptions about circuit and system function. We will outline the functional anatomy of the olfactory pathway in this manner here but then focus special attention on two frequently overlooked aspects of the circuit. First, we will discuss variation in odor sampling at the receptor sheet that can be created by both individual differences in receptor functional expression and by behaviorally induced variation in airflow patterns. Second, we will discuss the structure and potential function of cortical feedback circuits. In other sensory systems, these circuits (e.g., corticothalamic pathways) play critical roles in selective attention and expectation effects on sensory processing (e.g., priming effects). Do similar circuits and functions exist in olfaction?

Basic Olfactory Circuits

The basic architecture of the olfactory system is remarkably convergent across the animals, including humans. Our description of the mammalian system here is based largely on data obtained from rodents, but it seems to be equally relevant in humans. Thus, this section lays the anatomical foundation for discussions of function in both humans and animals in later chapters.

In its most basic form, the primary olfactory pathway traditionally involves a receptor sheet, either within the nasal passage of vertebrates or on antennae or specialized surface structures in many invertebrates, a first central relay generally called the olfactory bulb in vertebrates or antennal lobe in insects, and a third-order structure called olfactory cortex in mammals or mushroom bodies in insects. For mammals, this is a highly simplified schematic in that the olfactory cortex is actually composed of many subareas including the anterior olfactory nucleus (a.k.a. anterior olfactory cortex [Haberly 2001]), the olfactory tubercle, and the anterior and posterior piriform cortices. Each of these subregions of the olfactory cortex has a unique architecture, unique afferent and efferent connectivity, and presumably unique functional contributions to olfactory coding and perception. However, for the most part, the comparative sensory physiology of these different structures is seriously understudied with the recent exception of the piriform cortex. In addition to this direct pathway, the mammalian olfactory pathway is richly interconnected with other cortical and limbic structures, including the orbitofrontal, entorhinal and perirhinal cortices, the hippocampus, and the amygdala. Furthermore, as noted above, most of these structures not only receive olfactory input but also provide feedback to neurons in the primary olfactory pathway. Finally, extensive modulatory and multimodal inputs terminate throughout the olfactory pathway, allowing contextual and state-dependent processing of odorants starting very early in the pathway.

The olfactory receptor sheet of both vertebrates and invertebrates includes receptor neurons that express from dozens to hundreds of different genes that code for olfactory G-protein-coupled receptors (Buck 1996). In mammals, it appears that a single-receptor neuron expresses a single-receptor gene, though in invertebrates multiple gene expression within single-receptor neurons may be more common. The ligands for olfactory receptors, for the most part, are not entire odorant molecules, but rather submolecular components or features of odorant molecules (Araneda, Kini, and Firestein 2000). For example, an olfactory receptor neuron may be selectively excited by odorants containing a hydrocarbon chain of a particular length and/or a specific functional group in a certain location. Thus, given that most odors that we perceive are composed of multiple molecules and each molecule may be recognized by multiple receptor neurons each binding a unique feature of the molecule, odors will produce widespread and complex activation of a large population of different olfactory receptors. It should also be noted that odorants (or odorant features) can competitively interact at the receptor-binding site, adding further complexity to the spatiotemporal activity patterns leaving the receptor sheet.

The combination of receptors activated by a particular odorant is converted into a spatial pattern in the olfactory bulb (antennal lobe) through precise convergence of receptor neuron axons within olfactory bulb glomeruli. Each glomerulus appears to receive input from a homogeneous population of receptor neurons all expressing the same receptor gene. Because of the homogeneous receptor input to a glomerulus, odor-evoked activity within each glomerulus essentially reflects the binding specificity of the odorant receptors on the neurons projecting to that glomerulus. Based on functional imaging of odor-induced glomerular activity, receptor neurons transducing similar odorant features appear to target neighboring glomeruli. Thus, for example, glomeruli responding to aldehydes varying in carbon chain length all cluster near to each other within the olfactory bulb, whereas glomeruli responding to esters varying in carbon chain length cluster near to each other in a different region of the bulb (Leon and Johnson 2003). Spatial clustering of neurons encoding similar odorant features allows for lateral inhibitory circuits to enhance the contrast between these similar features. Although there are two levels of inhibitory interneurons within the olfactory bulb, juxtaglomerular neurons and granule cells, we hypothesize that juxtaglomerular neurons may play a more critical role in this type of lateral inhibition than granule cells (see below). Juxtaglomerular neurons not only allow lateral interactions between glomeruli, but also can directly regulate olfactory nerve input via presynaptic contacts.

The notion of a spatial map of odorant features laid out across the glomerular sheet is enticing, and quite convincing in its experimentally derived form (fig. 3.5). In fact, odorant identity can be predicted in some cases by observation of the complete odorant feature map in the rat olfactory bulb (Linster et al. 2001). Under natural conditions, however, odorants are rarely ex-

Fig. 3.5. The same high-contrast image shown in figure 3.3 is shown here in normal gray scale. Observing this image allows the visual system to learn how to group features in figure 3.3, such that now the macaque's face is apparent even in that highly degraded image.

perienced as isolated, pure stimuli appearing against a clean air background. Rather, on any given inhalation, many molecular species are sampled, some of which come from a single source, whereas others come from diffuse, perhaps background sources. What is perceived in this case cannot necessarily be predicted from the complex odorant feature map of the glomerular layer. Instead, the glomerular layer spatial map must be selectively read and interpreted by second- and third-order olfactory neurons.

In mammals, the principal second-order neurons, mitral and tufted cells, have single apical dendrites that receive input from a single glomerulus, and thus receive homogeneous receptor input. In the mammalian accessory olfactory bulb and main olfactory bulb of many nonmammalian vertebrates and invertebrates, mitral and tufted cells receive dendritic input from more than one glomerulus. The functional consequences of these two different patterns of connectivity have not been fully determined. Mitral and tufted cells in the mammalian main olfactory bulb are the projection neurons of the bulb and send axons throughout the olfactory cortex. Mitral and tufted cells appear to have somewhat different odorant response properties and projection patterns, though, again, the significance of these differences for odor behavior and perception have not been fully investigated (Scott and Harrison 1987; Nagayama et al. 2004).

Given the selective receptor input that individual mitral and tufted cells receive and the lateral inhibitory circuitry within the olfactory bulb, mitral and tufted cells seem to function largely as odorant feature detectors. Activity within the olfactory bulb thus refines the feature extraction process initiated at the receptor sheet and translates the largely spatial expression of odorant identity into a temporal pattern of spike trains to be read by third-order neurons in the cortex.* Olfactory bulb output neuron activity is not only driven by odorant input, but it is also shaped by extensive centrifugal inputs to the bulb, largely via interactions with intervening granule cells. Granule cells are axonless, GABAergic interneurons that outnumber mitral cells by about 50:1. Granule cells form dendrodendritic reciprocal synapses with mitral and tufted cell secondary dendrites and are hypothesized to mediate activity-dependent feedback and lateral inhibition of the output neurons, as well as entrain temporal patterning of output spike trains. However, as described below, granule cells are also the target of centrifugal feedback to the bulb and thus could play an important role in the effects of attention and expectancy in odor coding in the bulb.† Granule cells undergo experience-dependent neurogenesis and apoptosis throughout life, with enhanced survival of granule cells in regions encoding familiar or meaningful odors (Lledo and Gheusi 2003), further suggesting an important role in encoding familiar odors.

Odor-related activity within the olfactory bulb shows a strong oscillatory nature as originally observed even during some of the first electrophysiological recordings from the olfactory bulb (Adrian 1950). These oscillations can be observed in local field potentials and occur in both vertebrates and invertebrates. The temporal structure of these oscillations suggests at least three

---

* Recent data suggest that a temporal dimension to odorant coding exists even within the glomerular layer, where different glomeruli may have different temporal patterns of activation, perhaps depending on from where within the receptor sheet they receive input. It is not clear at present if these latency shifts contain critical information about odor identity, or rather if they are noise that then must be dealt with by subsequent processing.
† Thus, again arguing that odor perception cannot be predicted solely from the features of the glomerular layer spatial map.

origins in mammals, a high-frequency (40–90 Hz gamma frequency) local circuit oscillation driven by mitral cell–granule cell reciprocal synapses, a moderate-frequency (15–40 Hz beta frequency) oscillation driven by feedback loops between the olfactory bulb and cortex, and a low-frequency oscillation (2–15 Hz theta frequency) driven primarily by receptor input and reflecting the respiratory cycle. In mammals, these oscillations have also been hypothesized to reflect at least two different processes: those critical for odor identity and those reflecting top-down processes such as expectancy and attention (Freeman 1978; Kay, Lancaster, and Freeman 1996; Martin et al. 2004). Although the evidence that high-frequency oscillations reflect activity involved in expectancy is increasingly strong (e.g., olfactory bulb gamma-frequency oscillation emerges prior to odor onset in animals well trained in a discrimination task; Martin et al. 2004), their role in encoding odor identity appears minimal at best (Fletcher et al. 2005).

These local field potential oscillations, of course, are created by membrane currents of individual neurons, and as membrane current fluctuations become more synchronous within a given frequency range across a population of neurons, the power of local field potential oscillations at that frequency increases. Changes in synchronous activity of olfactory bulb output neurons are not only reflected as changes in the amplitude of local field potential oscillations, but should also affect the extent to which target cortical neurons receiving this input are driven. Thus, simple coincidence detection by olfactory cortical neurons would allow single cortical pyramidal cells to respond to unique combinations of mitral cell (and thus olfactory receptor) input. In this way, the olfactory cortex could synthesize the diverse odorant feature extracted at a particular point in time by the periphery into unique, whole odors.

However, the mammalian piriform cortex (fig. 3.7) goes far beyond simple, passive coincidence detection to synthesize odorant features into odor perceptual objects. Termination of mitral cells conveying information from olfactory receptors expressing one type of receptor gene overlaps with input from other receptor types, allowing for convergence of multiple receptors/odorant features onto single cortical neurons (Zou et al. 2001). This convergence is dramatically enhanced by an elaborate network of intracortical association fibers, which terminate on the proximal apical dendrites and basal dendrites of Layer II/III pyramidal cells. The mitral/tufted cell input and association fiber systems, in addition to being anatomically segregated on the cortical neuron dendritic tree, also appear to have significant differences in physiology, including plasticity and modulatory control. For example, affer-

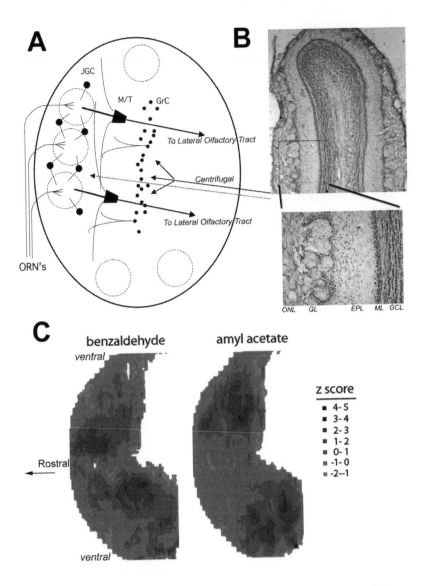

Fig. 3.6. The basic organization of the mammalian olfactory bulb. (A) Axons of olfactory receptor neurons (ORNs) terminate in olfactory bulb glomeruli, which are surrounded by the cell bodies of juxtaglomerular neurons. Each glomerular receives input from a homogeneous population of ORNs expressing the same receptor gene. Within the glomeruli, ORNs make excitatory synapses with juxtaglomerular and mitral/tufted cells (M/T). M/T cells are the output neurons of the olfactory bulb and their axons form the lateral olfactory tract (LOT). Granule cells (GrC) for reciprocal dendrodendritic synapses with the lateral dendrites of M/T cells, and are also the primary target of most centrifugal input to the olfactory bulb. (B) A cresyl violet-stained section through the

ent synapses display activity-dependent short-term depression that may contribute to habituation and adaptation to background odors (Best and Wilson 2004). This rapid cortical adaptation results in a system most effectively driven by changing stimuli—static background odors can be filtered while novel odors or odors fluctuating in intensity continue to receive cortical attention (Kadohisa and Wilson, forthcoming). This simple mechanism of cortical adaptation could allow parts of the olfactory bulb spatial map to be read, while other parts are filtered (Wilson 2004).

In contrast to the afferent synapses, association fiber synapses display robust associative long-term plasticity that may be critical for storing records of previously experienced patterns of odor feature input (De Rosa and Hasselmo 2000; Haberly 2001). Anatomically and physiologically realistic computational models of piriform cortex suggest autoassociative memory capabilities (Hasselmo et al. 1990; Haberly 2001). Autoassociative memory circuits involve a distributed excitatory input and extensively distributed excitatory intrinsic association fibers—characteristics expressed by the piriform cortex. Furthermore, the intrinsic association fibers should be capable of expressing Hebbian synaptic plasticity—again, a characteristic of the piriform cortex. This type of circuit extracts and learns correlations between patterns of afferent input by strengthening intrinsic connections between coactive neurons.

This learning has two consequences. First, repeated patterns of afferent input result in reproducible and stable cortical ensemble representations and, in turn, cortical output. Those neurons that have been repeatedly coactive strengthen their common excitatory connections and become a more reliably evoked ensemble in response to a given afferent input. It follows that stable cortical output should contribute to stable odor perceptions over repeated stimulation with the same familiar stimulus. Disruption of intrinsic synapse plasticity leads to variable cortical representations in computational models (Hasselmo et al. 1990), and disrupted cortical (Wilson 2001) and behavioral (De Rosa and Hasselmo 2000; Fletcher and Wilson 2002) odor discrimination.

---

rat olfactory bulb and a higher magnification of olfactory bulb lamina (ONL, olfactory nerve layer; GL, glomerular layer; EPL, external plexiform layer; ML, mitral cell body layer; GCL, granule cell layer). (C) Examples of odor-specific spatial patterns of glomerular activity mapped with ^{14}C-2-deoxyglucose. The olfactory glomerular layer is viewed as if peeled off of the olfactory bulb and laid flat with the ventral side of the bulb now split and lying on the top and bottom of each image. These two different odorants evoke unique, though partially overlapping patterns of activity (^{14}C-2-deoxyglucose maps provided by Brett Johnson and Michael Leon, University of California at Irvine).

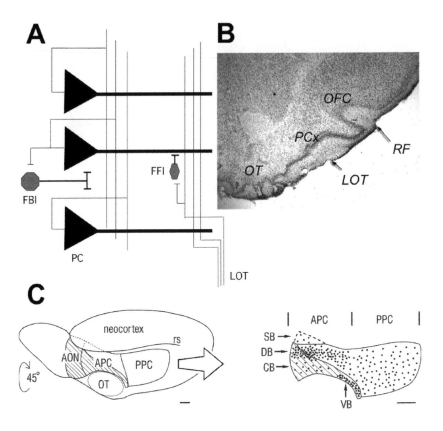

Fig. 3.7. The basic organization of the mammalian piriform cortex. (A) Pyramidal cell (PC) dendrites are the target for both lateral olfactory tract (LOT) input from the olfactory bulb (terminating on the distal half of the apical dendrite) and intracortical and commissural association fibers (terminating on the proximal half of the apical dendrite). Inhibitory interneurons include superficial feedforward inhibitory (FFI) and deeper feedback inhibitory (FBI) interneurons. (B) A cresyl violet-stained section through the rat olfactory cortex showing olfactory tubercle (OT), piriform cortex (PCx), and orbitofrontal cortex (OFC). The rhinal fissure (RF) and LOT are also labeled. (C) A stylized map of c-*fos* immunohistochemical labeling of odorant-activated neurons within the piriform cortex. Each dot represents cell labeling. Although there is differential cell labeling in response to odorants across subdivisions of the piriform cortex, current evidence suggests no or minimal odor specific activity patterns, i.e., different odorants evoked roughly the same spatial pattern (c-*fos* immunohistochemistry image provided by Kurt Illig, University of Virginia).

The second consequence of cortical autoassociative function is pattern completion in the face of degraded or slightly modified patterns of afferent input. Computational models suggest that once a pattern has been learned by the piriform cortex (or any autoassociative network), complete representations can be expressed in the output even if the input is missing a subset of features. Upon subsequent stimulation with familiar patterns of odorant features, the cortex rapidly recognizes the odor, even if the input is partially degraded. This robust cortical pattern recognition in the face of partial inputs may be particularly important during odor intensity variation, which can produce changes in receptor activity, and/or in situations where some features may overlap with background features and have thus been adapted out. The experience-dependent changes in piriform odor encoding may contribute to improvement in odor discrimination acuity (olfactory perceptual learning) that occurs with odor familiarization (e.g. Fletcher and Wilson 2002).

The highly associative piriform cortical network is hypothesized to be critical for the final synthesis of the multiple odorant features extracted and refined peripherally into unique odor perceptual objects (Haberly 2001).* However, the piriform cortex may also play a role in the multimodal, contextual nature of odor perceptions. Neurons in the piriform cortex respond not only to specific odors, but also to odor-contextual cues (Schoenbaum and Eichenbaum 1995; Critchley and Rolls 1996; Ramus and Eichenbaum 2000).

Neurons in the olfactory bulb and the piriform cortex also project directly to the amygdala and entorhinal cortex, the latter of which serves as the sensory gateway into the hippocampal formation. The role of the hippocampal formation and the amygdala in memory for olfactory cues has been well described (Staubli, Le, and Lynch 1995; Eichenbaum 1998) and will not be described in detail here (see chapter 6). Heavy innervation of the amygdala by primary olfactory structures, however, provides a powerful mechanism for the rich experience that can stem from olfactory sensation. The role of the amygdala in emotion, memory, and autonomic control directly ties olfaction to these primordial functions and adds complexity to the odor perceptual experience.

Finally, as noted above, the olfactory pathway is heavily innervated by

---

* In fact, Haberly (2001) hypothesizes that some or much of the initial odorant feature synthesis may occur within the anterior olfactory cortex and that piriform cortex is a site for multimodal synthesis. The anterior olfactory cortex is a seriously understudied region of the olfactory system. Until data exist for this region, we will emphasize the piriform cortex as the site of synthesis.

modulatory systems known to regulate plasticity and memory, which also depend on attention and arousal. Both the olfactory bulb and cortex receive a
strong cholinergic input from the horizontal limb of the diagonal band of
Broca (HLDB), which itself is responsive to olfactory input (Linster and Hasselmo 2000). This creates an interesting feedback loop in which cholinergic
modulation of olfactory processing is itself partially under olfactory control.
Norepinephrine from the nucleus locus coeruleus also heavily innervates the
olfactory bulb and cortex (Shipley and Ennis 1996). Norepinephrine release
in the olfactory bulb depends on the behavioral state and multimodal stimulus novelty. In the olfactory system, norepinephrine modulates both mitral/
tufted and piriform cortical neuron response to odors as well as short-term
and long-term plasticity (Wilson, Best, and Sullivan 2004).

Thalamocortical Olfaction

Although one of the unusual characteristics of the mammalian olfactory system is the lack of a thalamic relay between initial central structures and the
primary sensory cortex, olfactory information does have a thalamic representation and a thalamocortical projection. The mediodorsal nucleus of the thalamus receives direct projections from the piriform cortex and, in turn, projects to the orbitofrontal cortex (and other prefrontal regions). In addition to
thalamic input, the orbitofrontal cortex also receives direct input from the
piriform cortex (Johnson et al. 2000), and the topography of these two inputs
suggests a convergent triangulation between piriform, mediodorsal thalamus,
and orbitofrontal cortex (Ray and Price 1992). The mediodorsal thalamus also
receives input from the cortical and basal nuclei of the amygdala, potentially
enriching the olfactory input to the orbitofrontal cortex with emotional context (Schoenbaum, Setlow, and Ramus 2003). As in the piriform cortex, orbitofrontal neurons respond not only to specific odors, but also to odor-contextual
cues (Schoenbaum and Eichenbaum 1995; Critchley and Rolls 1996; Ramus
and Eichenbaum 2000). Neurons in the orbitofrontal cortex also respond to
specific odor-taste compounds (de Araujo et al. 2003).

The specific functions of thalamocortical olfactory pathways in odor perception are not known. Lesions of the dorsomedial thalamus in rats impair
learning and memory of discriminative odor cues (Slotnick and Kaneko 1981),
although this impairment appears to not be related to odor discrimination per
se (Staubli, Schottler, and Nejat-Bina 1987; Zhang et al. 1998).

In the orbitofrontal cortex of both rodents and primates, single neurons encode both odor quality and odor associations. In primates, both the identity of an odor being inhaled and its reinforcement value and taste associations can be extracted from orbitofrontal cortex neuron spike trains (Rolls, Critchley, and Treves 1996). Odor quality is encoded very sparsely, and there appears to be diversity in responses, with for example some neurons encoding odor quality independently of reinforcement associations and some encoding the reinforcement associations (Critchley and Rolls 1996). Similar results have been seen in the rat orbitofrontal cortex in different experimental conditions, and have led to the suggestion that the orbitofrontal cortex integrates sensory representations with associated reward value to help guide motivated behavior (Schoenbaum, Setlow, and Ramus 2003). This interpretation from single-unit recordings is supported by lesion studies (Otto and Eichenbaum 1992; Schoenbaum, Setlow, Saddoris, and Gallagher 2003).

Functional imaging studies in humans further support the notion of orbitofrontal cortex as an integrator of sensory input with meaning and reward value. Data suggest a differential encoding of inherently pleasant and unpleasant, and appetitively and aversively reinforced odors in the human orbitofrontal cortex (Gottfried, O'Doherty, and Dolan 2002). Furthermore the orbitofrontal cortex is differentially activated during active decision-making about odor pleasantness and unpleasantness (Royet et al. 2003), and can even be affected by concurrent visual stimulation (Gottfried and Dolan 2003). The strong connectivity of the orbitofrontal cortex with the amygdala (described above) may contribute to odor association encoding in this region (Savic 2002; Anderson et al. 2003). Finally, there is increasing evidence for lateralization in human thalamocortical olfaction, with the right hemisphere dominating (Zatorre et al. 1992), though lateralized differences in nasal odor sensitivity may contribute to this effect (Sobel et al. 1999).

From this description of the basic olfactory pathway several points can be extracted. First, the initial stages of the olfactory system, including the receptor sheet and olfactory bulb, are highly analytical, extracting and refining submolecular features from complex odorants. This is apparent from the odor response properties of individual receptor neurons, olfactory glomeruli, and mitral/tufted cells. Odorant features may interact with each other, adding complexity and uncertainty as to what defines an odorant feature, but nonetheless these early stages are highly analytical. The second point is that even at the level of second-order neurons in the olfactory bulb, encoding odorant features occurs in the context of other external and internal events. This point will be examined further below. Third, the synthesis of odorant features into perceptual odor objects involves more than simple detection of coincidence in mitral/tufted cell output by third-order neurons. Rather, experience-dependent cortical processes learn previous patterns of coactive feature input and match the current input to these representations stored in modified synapses of piriform cortical association fibers. The learning of familiar feature inputs allows recognition of distinct odors even when inputs are degraded by adaptation or interference. The learned cortical representation may also include multimodal or contextual components. Fourth, the overlap of olfactory and gustatory pathways in the orbitofrontal cortex and the heavy innervation of the amygdala by olfactory structures provides privileged interaction between odors and tastes and between odors and emotional and autonomic responses. This can lead to synesthetic perceptual qualities such as sweet odors and the robust emotional qualities that odors often evoke. That is, odor percepts, rather than being highly analytical as one might imagine from the early stages of the olfactory pathway, are instead highly synthetic and multimodal, driven not only by the odorant features extracted by the receptors but also by the context and experience and expectations of the smeller. Thus, as noted above, odor percepts cannot be predicted from the spatial maps of odorant features displayed across the olfactory bulb glomerular layer. Rather, although these maps place constraints on the odor percept, experience, expectation, and attention combine with odor input to result in the final perceptual experience.

Variation in Stimulus Sampling and Transduction

In mammals and many other vertebrates, odors initially gain access to the olfactory receptor sheet as a by-product of respiration. Odor molecules are inhaled during each respiratory cycle, or reach the receptor sheet retronasally from food in the mouth. Pinching the nostrils shut when eating prevents retronasal airflow and eliminates the olfactory components of flavor. In addition to these passive stimulus acquisition processes, odors can be actively sampled through sniffing. In many animals, active sniffing is a highly stereotyped behavior involving rapid, repeated (5–12 Hz theta frequency) short inhalations and is expressed during exploration, behavioral arousal, or exposure to novel stimuli. In humans, active sniffing appears to be more variable in temporal structure but nonetheless involves a change in rate and depth of inhalation during intentional sampling of the odor environment.

Active sniffing has at least two consequences, one mediated at the receptor sheet itself and one involving more central processes (also see chapter 4).

Peripherally, active sniffing changes airflow dynamics within the nasal passages. Changes in airflow may modify accessibility of odorants to different regions of the receptor epithelium. These airflow changes could increase the epithelial surface area activated by odorants, thus activating more receptors, in general, or potentially leading to activation of different receptor subpopulations on isolated nasal turbinate spaces. In addition, sniffing could modify the temporal structure of receptor activation, in turn, modifying activation of glomerular circuits. The temporal structure of receptor input to the olfactory system has been shown to be critical to olfactory bulb processing of odor identity in insects (Heinbockel, Christensen, and Hildebrand 1999).

Centrally, changes in the frequency of inhalation-driven afferent volleys to the olfactory system could have a number of consequences for odor coding. For example, the theta frequency of active sniffing is a particularly effective method of activating N-methyl-D-aspartate (NMDA) receptor-dependent synaptic plasticity (Larson, Wong, and Lynch 1986). In addition to regulating long-term synaptic plasticity, NMDA receptors are also involved in normal synaptic transmission throughout the olfactory system, including the olfactory nerve to mitral cell synapse, the mitral cell–granule cell synapse, and afferent and association fiber synapses within the olfactory cortex. Changes in the rate of synaptic activation can modulate the extent of NMDA receptor activation, producing, in essence, a sniffing-induced regulation of synaptic gain and/or inhibitory control (Young and Wilson 1999).

In addition to sniffing, recent work suggests that olfactory stimuli may induce olfactomotor reflexes that subtly and rapidly modulate inhalation volume depending on odor intensity (Johnson, Mainland, and Sobel 2003). Sobel and colleagues have argued that these rapid adjustments in inhalation volume could contribute to perceptual constancy over changes in stimulus intensity by modulating the amount of odorant inhaled (Johnson, Mainland, and Sobel 2003). Similar peripheral mechanisms of reflexive adjustments in stimulus sampling with changes in intensity occur in other sensory systems, for example, pupil dilation and constriction in vision and motor control of middle ear sound amplification as sound intensity increases.

Finally, once the odorant has reached the receptor sheet, variations in receptor expression or activity between and within individuals can further influence how the stimulus is sampled. Thus, for example, genetic differences between individuals could affect receptor expression, potentially resulting in specific anosmias in the most extreme case or influencing relative sensitivity to specific odorant features (Wysocki and Beauchamp 1984; Wysocki, Dorries, and Beauchamp 1989). Although specific anosmias have been described in the literature, given what we currently believe about combinatorial odorant coding at the receptor level, it should be extremely rare to have a tight correlation between loss of a specific receptor and an inability to smell a particular odor.

More commonly, rather than a complete loss of ability to smell particular odors individuals often express differences in relative sensitivity to odors. These differences in relative sensitivity may stem from either differences at the receptor sheet or from central differences. Variability in receptor sensitivity could be due to genetic differences in receptor expression, experience-dependent differences due to odorant exposure (or lack of exposure) during development or after maturation, or differences in the perireceptor environment such as hormonal or mucosal changes. Experience has been demonstrated to affect sensitivity and/or affective responses to odors in both neonatal and mature animals and humans, though the peripheral or central locus underlying these changes in sensitivity have not been fully identified (Wysocki, Dorries, and Beauchamp 1989; Mainland et al. 2002). Similarly, hormonal and autonomic regulation of the olfactory epithelium and receptor neurons have been described that could be mediated within individual variation in odorant feature sensitivity (Chen et al. 1993; Eisthen et al. 2000). These latter changes may be expected to have a broader impact on olfactory sensitivity, as they would be expected to affect the complete or at least a wider subset of the receptor population.

Together, these behavioral, physiological, and genetic differences in odorant sampling and transduction lead to differences in odor perception. Knowledge of the physicochemical stimulus alone is not sufficient, therefore, to always predict the resulting perception either across different individuals or in some cases even within individuals over time. These kinds of stimulus-sampling issues have proven to be critical factors in recent functional imaging studies in humans, where precise instructions and stimulus control are required to evoke consistent patterns of odor evoked brain activation.

Feedback Circuits

As exemplified in our description of the visual system above, feedback from more central stages to more peripheral stages of sensory pathways is a basic component of thalamocortical sensory systems. These feedback circuits serve a variety of functions including gain control, selective attention, and priming or expectancy-based tuning. As we have attempted to clarify, perceptual objects are not inherently obvious from the information extracted by the sensory receptor sheet in either the visual or olfactory systems. Thus, simple feedforward information flow may be insufficient to account for object perception, perceptual constancy, and scene analysis. Feedback circuits allow matching of partial input patterns with previously experienced (and memorized) patterns to allow pattern completion and subsequent object recognition. Feedback pathways also allow introduction of expectancy and selective attention to allow or facilitate transmission of some patterns through to cortex and gating or suppression of other irrelevant or unexpected patterns. This feedback need not come solely from the directly superordinate stage, but also from neurons several stages removed or even from nonsensory structures such as the hippocampus.

The olfactory system is a superb example of feedback loops, yet although these circuits have been more or less well described anatomically, with a few exceptions the function of feedback in olfactory processing and odor perception remains relatively unexamined (Grajski and Freeman 1989; Kay, Lancaster, and Freeman 1996). One of the unique aspects of the olfactory pathway, as noted above, is that the receptor neurons synapse directly within a cortical structure, the olfactory bulb, rather than a more simple brainstem nucleus or the relatively isolated retina. The olfactory bulb receives massive feedback from all of its projection targets, including the anterior olfactory nucleus, piriform cortex, and amygdala. This feedback is in addition to modulatory inputs from the locus coeruleus, the horizontal limb of the diagonal band, and the raphe nucleus. The majority of this feedback terminates on inhibitory interneurons in the granule cell layer, with some additional feedback directly to juxtaglomerular neurons.

Modulatory inputs to the olfactory bulb have been shown to be involved in state-dependent control of sensory evoked responses, mitral cell adaptation to odors, and experience-dependent plasticity, as mentioned above. The specific role of cortical feedback on olfactory bulb function and odor perception is less clear. However, using analyses of local field potentials recorded in the olfactory bulb, Freeman and colleagues have proposed that this feedback, as measured with spatial and temporal patterns of gamma-frequency oscillations, reflect expectancy and/or attentional processes feedback to the olfactory bulb from cortical or limbic structures. Thus, for example, gamma-frequency oscillations can be observed to increase in the olfactory bulb prior to the onset of odor sampling in well-trained rats, and then decrease during actual stimulus sampling (Kay, Lancaster, and Freeman 1996; Martin et al. 2004). Furthermore, using a 64-electrode array for sampling spatial patterns of olfactory bulb local field potentials, Freeman and Schneider (1982) demonstrated that spatial patterns of gamma frequency olfactory bulb activity did not reflect odor identity, but rather most strongly reflected past experience and expectancy. That is, training an animal that a particular odorant (e.g., odor A) signaled electric shock resulted in a unique spatial pattern of olfactory bulb gamma-frequency oscillations. After the animal has learned odor A, presentation of odor B induced the same spatial pattern of activity as odor A. Only if the animal was now trained with odor B did the spatial pattern shift to a new pattern, which was then stable for all odors tested.

Based on these and other results, Freeman suggested that the spatial patterns of olfactory bulb gamma-frequency oscillations did not reflect the identity of the odor being inhaled, but rather reflected a template of the expected or recently learned odor which was imposed on the bulb from cortical feedback. This feedback could presumably enhance recognition of the odor if it matched the expected pattern of the template. If the odor did not match the imposed template, the odor may be misidentified or the template altered if conditions warranted (e.g., the new odor was now paired with shock).

Cortical feedback thus could serve as an expectancy or attentional generated searchlight, facilitating identification of familiar odorants and initially confusing or filtering information about novel odorants. As those novel odorants become familiar or gain learned significance through changes in cortical circuits, or as perhaps multimodal or contextual cues dictate (e.g., "this is a red wine and thus should have these olfactory components"), the patterns of cortical feedback change.

This view of cortical feedback is highly reminiscent of corticothalamic feedback in the visual system described above. Recall that in vision (and other thalamocortical systems), reticular thalamic neurons receive excitatory input from the sensory thalamus and descending excitatory input from the sensory cortex. These reticular thalamic neurons, in turn, provide an inhibitory input to the sensory thalamus that can gate, through inhibition and disinhibition, subsequent sensory flow back to cortex. This feedback-controlled inhibition, therefore, serves as a searchlight to select which information will be transmitted to the cortex based on expectancy and attentional needs (Crick 1984; McAlonan and Brown 2002).

It is tempting to speculate that granule cell interneurons play a role in olfaction comparable to that of the reticular thalamus in vision and other thalamocortical sensory systems. Granule cells monitor mitral cell output via dendrodendritic synapses and receive axosomatic and axodendritic synapses from cortical feedback fibers (Price and Powell 1970). Granule cells also receive inhibitory synaptic input from olfactory bulb intrinsic neurons (Price and Powell 1970). Although several hypotheses exist regarding the role of mitral-granule cell reciprocal connections in lateral and feedback inhibition, these hypotheses largely overlook the impact of excitatory cortical input to granule cells. Details of termination patterns of individual cortical neurons projecting to the bulb are not yet available. However, olfactory cortical feedback to the bulb could provide a higher-order modulation of granule cell excitability, resulting in top-down control of how sensory afferent activity is processed by the bulb. This cortical modulation of granule cell excitability could be evidenced by changes in gamma-frequency local field potential oscillations during periods of strong descending input to the bulb, just as has been found in freely behaving animals performing well-learned olfactory behaviors, as described above. As described by Freeman many years ago (Freeman and Schneider 1982), and observed since (Kay and Laurent 1999), through such descending control, attention and expectancy can modulate how the olfactory bulb processes odors. The clear implication is that odor perception is not simply driven by the physicochemical stimuli entering the nose, but also by context, behavioral state, expectations, and memory.

Summary

Olfactory perception involves information processing far beyond simple molecular feature detection occurring at the receptor sheet. Olfactory perception can be more accurately described as an object-oriented process, demonstrating characteristics of perceptual constancy and scene analysis, and is affected by expectancy, attention, and experience. Olfactory stimulus sampling and central olfactory processing circuits provide mechanisms for each of these characters. Behavioral regulation of stimulus sampling through active sniffing and movement of the nose can impart attentional and spatiotemporal components to an olfactory scene. Spatial maps of odorant features allow lateral inhibition to enhance contrast between similar odorant features and fine-tune loose input from broadly tuned receptors. Associative circuitry within the olfactory cortex and synaptic plasticity allow object synthesis. Multimodal convergence and strong limbic system associations can enhance the richness of olfactory object percepts. Finally, extensive feedback circuitry throughout the olfactory cortex and strongly innervating the granule cell layer of the olfactory bulb can result in odor perception shaped by attention, expectancy, and past experience.

4

The Relationship between Stimulus
Intensity and Perceptual Quality

This chapter concerns innate constraints on odor perception, in particular, in the context of the effects of stimulus intensity on perceptual quality. A discussion of the effects of intensity on perceptual quality highlights the fact that odor quality does not solely depend on stimulus structure and sets the stage for our subsequent argument regarding learned odor objects. The first part of the chapter examines issues related to the concentration of stimuli, namely detection at threshold, and suprathreshold perception of intensity. Historically, individual variation in the limits of detection for different stimuli has been seen as a route to understanding the physiological and ultimately the psychological basis of olfaction (Amoore 1970; Beets 1978). This approach is analogous to the study of color blindness and its role in understanding the basis of color vision (see Boring 1950). Although individual odor thresholds do vary, more recent experimental work indicates that variation within an individual may in fact be of comparable magnitude over time to variation between individuals at one time (Stevens, Cain, and Burke 1988). Nonetheless, some individual differences in threshold are probably initially under genetic control, but even here, experience can act to up-regulate the expression of receptors and may alter how information is processed centrally too. This is followed by two further findings related to variation in suprathreshold responses—the effect of familiarity on judgments of intensity (Hudson and Distel 2002) and perceptual constancy of intensive judgments (Teghtsoonian et al. 1978). In both these cases, a basic psychophysical parameter, intensity, can be affected by purely psychological variables and, in the former case, solely by experiential factors. The final issue in this section concerns changes in the quality of an odorant, as its concentration is increased (Gross-Isseroff and Lancet 1988). This is a key point, because it suggests that chemical structure alone cannot dictate quality, as structure remains constant under conditions of high or low concentration. Rather perception must rely on the interpretation of patterns of receptor output, as these may change due to the higher likelihood of a chemical binding to new receptors as concentration increases.

Odorant Concentration

Individual Variation in the Ability to Detect Odorants

Identifying individuals with specific deficits in their ability to detect certain odorants has been an area of intense interest for the past 50 years (Guillot 1948). Initially, the aim was to collect information that might reveal relationships between specific physicochemical classes of odorant and specific receptors (Amoore 1970; Beets 1978). If, for example, an individual lacked a receptor capable of detecting a certain type of structure, then one might expect a specific anosmia for odorants of that general class. One proponent of this approach was Amoore, who collected data for more than three decades revealing over 76 different specific anosmias (see Amoore [1982] for review). For the most part, these anosmias were not absolute inabilities to smell these compounds, although, as we discuss below, there may be some chemicals that do fit this description. Instead, these anosmias tended to be far less able to detect them at threshold than control participants, while showing comparable thresholds for unrelated stimuli. For example, Amoore and Forrester (1976) explored thresholds for trimethylamine and related amines. The 286 normosmics for trimethylamine could detect it at a threshold 2290 times below that of the specific anosmic group ($n = 21$). Similarly, the anosmics were also uniformly poorer at detecting other amines.

Such effects do not appear to result from general anosmia. Baydar, Petrzilka, and Schott (1993) found that there was no correlation between thresholds for androstenone and galaxolide, that is, performance in detecting one compound could not predict performance in detecting the other. Specific anosmias for androstenone have been explored in considerable depth, in part, because androstenone may be a human pheromone. There is evidence that in adults, ability to detect androstenone is significantly poorer in men than in women, but this gender difference does not manifest in prepubescent children (Dorries et al. 1989). Adult men excrete androstenone in their sweat and urine to a far greater extent than women do, and there is some evidence that this chemical may cause a variety of sex-related behaviors in women (Hays 2003). Consequently, specific anosmia to androstenone has attracted considerable interest.

One puzzling observation was that exposure to androstenone in human males could result in a participant moving from being a nondetector to a detector. That is, exposure can result in an enhanced ability to detect androstenone (Wang, Chen, and Jacob 2004; Wysocki, Dorries, and Beauchamp 1989). Such effects are not limited to androstenone. Dalton and Wysocki (1996) observed a similar effect for two other odorants, citralva and isoborneal, i.e., exposure led to elevated sensitivity at threshold. This effect was not a general consequence of practice at detecting thresholds, because the change for the exposed odorant significantly exceeded that for the control. Although such effects may be a feature of olfaction, in general, the only studies exploring the locus of this effect have been with androstenone.

The data now available suggest that these exposure-mediated enhancements of sensitivity involve both a central and a peripheral component. Evidence for the latter comes from two studies. Yee and Wysocki (2001) found that exposure to either amyl acetate or androstenone in mice led to enhanced sensitivity at threshold. In a second experiment they established thresholds for one of these chemicals and this was followed either by a sham lesion of the olfactory nerve or an actual lesion. All the mice were then exposed to either amyl acetate or androstenone over a 10-day period. Because this exposure could only affect the olfactory epithelium in the lesioned animals because of the severing of the olfactory nerve, any subsequent benefit to threshold must presumably result from a peripheral (olfactory epithelium) cause. Consistent with this account, they found that following recovery from the lesion or sham surgery, all the mice had enhanced sensitivity for the exposed odor, relative to performance for the nonexposed odor. Similarly, in humans, evidence has also been obtained of a role for peripheral receptors in enhancing threshold detection. Wang, Chen, and Jacob (2004), found that olfactory evoked potentials (OEPs) recorded in the nose are correlated with detection thresholds, and that OEPs increased in line with the enhanced ability to detect androstenone, suggesting a peripheral site for this effect.

Whether the human data point to changes in receptor density or receptor sensitivity or are centrally mediated as well is unclear. But results from another recent study suggest that central factors must have some role in changing responsiveness. Mainland et al. (2002) found that exposure in humans to androstenone in one nostril, while the other was occluded, resulted in an enhanced threshold in the nonexposed nostril when tested after the exposure period had been completed. This result suggests that although peripheral factors may mediate detection effects, central factors must influence this, because there is no direct connection between the left and right olfactory epithelia.

We now know that some of the variability in thresholds attributed to "specific anosmia" probably results from other causes. First, there is a large degree of individual variation in sensitivity, which is of the same magnitude as that seen between individuals on one test session (Stevens, Cain, and Burke 1988). Second, traditional techniques of anosmia classification tend to overestimate the number of anosmics (Bremmer et al. 2003), although some people clearly still have deficits in the detectability for certain odorants. Moreover, some of this deficit can be eliminated by exposure to the target odorant, resulting in a heightened sensitivity for that compound, an effect that appears to be mediated at least in part by peripheral mechanisms (see above). Whether these changes in sensitivity enhance discriminability is not currently known; nonetheless, this form of plasticity is probably complementary to those described later in this book. The observation that most of these single compounds can be smelled at suprathreshold by both anosmics and normosmics (with the possible exception of androstenone), however, suggests again that the olfactory system does not generally rely on the presence of just one receptor type to detect a particular odorant.

Individual Variation in Suprathreshold Perception of Intensity

To our knowledge, the studies of androstenone discussed above have not explored whether shifts in threshold translate into suprathreshold changes in perceived intensity, but at least one study using this type of methodology (Dalton and Wysocki 1996) found that, although exposure enhanced sensitivity for citralva and iso-borneal, it had no effect on suprathreshold ratings of intensity. Several studies, however, have suggested that exposure may exert an influence at the suprathreshold level. The basic design here has been to compare participants' intensity ratings for odorants likely to be either familiar or unfamiliar to them (Hudson and Distel 2002). Ayabe-Kanamura et al. (1998)

compared forty Japanese and forty German participants on eighteen odorants, six of which were judged likely to be familiar to Japanese participants, six familiar to the Germans, and the remaining six likely to have been experienced by both groups. Two findings emerged. First, there was a clear tendency for odors to be judged as stronger smelling if they were familiar (i.e., culturally appropriate) to the participant. Second, however, was the observation that misidentification influenced this relationship, in that certain odors, such as dried fish for example, were judged highly familiar by German participants, but only because they were misidentified as feces. In this case intensity ratings apparently were also influenced by the hedonic reaction that this label evoked.

Distel et al. (1999) extended this work by testing a further sample of forty Mexican participants using the same design. They then examined the relationships by odor type and by individual, between intensity, hedonic, and familiarity ratings. As might be expected, they observed a positive and significant relationship between familiarity and intensity (mode = 0.4), irrespective of whether this was calculated across odors or within individuals. This suggests that the more familiar an odor is, the more likely it is to be rated as stronger smelling. Similar positive relationships were observed between absolute hedonic ratings and intensity, in that the more intense the odor was judged as smelling, the more pronounced the reaction (irrespective of sign). As demonstrated in many previous studies, positive relationships were observed for familiarity and hedonic ratings, in that familiar odors are generally judged as more pleasant smelling (e.g., Rabin and Cain 1989). Obviously this does not hold for all familiar odors.

Distel and Hudson (2001) tested the influence of familiarity on the judgment of intensity more directly by using a culturally homogeneous sample of German students. Each smelled two sets of odors under different conditions. The first condition involved presenting the odor, rating its intensity, then rating its pleasantness and familiarity, and finally, attempting to generate a name for it. In the second condition, participants were told the name of each of the odors before making the same three ratings. They were then asked to judge how well the name provided by the experimenters fit the odor they had smelled. Three important findings emerged. First, when a name was present, participants rated the odors as more intense, more familiar, and more pleasant. Second, under the first condition, comparing odors that could be named with those that could not revealed an identical pattern—greater intensity, familiarity, and liking for the named odor. Third, under the second condition, comparing odors that were judged to fit well with the name provided versus those judged to be a poor fit, those with a good fit were judged as more intense, familiar, and pleasant. Yet again, these findings and the others above, suggest that where an odor has been previously experienced, it will likely be judged to smell stronger.

An obvious question is whether exposure to an odorant under laboratory conditions can result in changes in intensity. Although this has never been tested directly, there is evidence that this can occur. Stevenson, Prescott, and Boakes (1995) and Stevenson, Boakes, and Prescott (1998) exposed participants to odor-taste pairings to study changes in judgments of sweetness and sourness, which are described later in this book. Intensity ratings were also obtained. Although the only consistent significant finding for intensity was a positive change for the sucrose-paired odor (that is this odor was judged more intense after it had been experienced several times in solution with sucrose) the same general trend was apparent in water-paired and citric-acid-paired odors too. The odors tended to be judged as more intense postexposure. It is not currently known whether this plasticity in the judgments of intensity seen in these experiments and reported in the others described above result from peripheral or central mechanisms. Nonetheless, they suggest that the olfactory system becomes sensitized to odors to which it is regularly exposed and that this can potentially manifest as an increase in perceived intensity and/or an increase in a participant's sensitivity to detect it, as described in the preceding section.

PERCEPTUAL CONSTANCY OF INTENSITY

In the visual system, distance cues serve to mediate size constancy, such that even though two objects may differ in the area of the retina on which they impinge, the brain can still correctly identify the larger object (Gregory 1966; see chapter 2). There is some evidence favoring a similar mechanism in olfaction. In this case sniff vigor is a key variable that allows the olfactory system to modulate the perception of intensity. It is well established in animals that increased flow rate of an odorant across the olfactory mucosa increases the level of activity in the olfactory nerve (Tucker 1963). Presuming this to be true in humans, one might presume that perceived intensity would also increase with flow rate. Teghtsoonian et al. (1978) investigated this in human participants by asking them to sample odors (amyl acetate and butanol) at several concentrations, using either a weak or a vigorous sniff for each concentration, both of the same duration. They found that magnitude estimates of perceived intensity did not change by sniff type, even though the quantity of odorant deposited on the olfactory epithelium must have been significantly greater in the vigorous sniff condition.

This result was explained by feedback provided by the sniff. That is when the participant engaged in a vigorous sniff, this information was used to calibrate the level of activity recorded by the olfactory epithelium to a level equivalent to that produced by a weak sniff. Support for this account was obtained in a further series of studies that independently varied sniff vigor and flow rate (Teghtsoonian and Teghtsoonian 1984), the latter being manipulated by changes in resistance of the inspired air. In this study, when sniff vigor was held constant, but flow rate varied, participants judged an odorant as more intense as flow rate increased. Conversely, when flow rate was constant, but sniff vigor changed, there was no accompanying alteration in intensity. These findings are consistent with a further series of studies by Teghtsoonian and Teghtsoonian (1982), which observed that changes in flow rate, produced by varying the resistance of the inspired air, did not alter ratings of judged sniffing effort, whereas changes in sniff vigor did.

Not all studies have supported these findings. Laing (1983) found that a high volume (or vigorous) sniff did produce higher-intensity ratings for cyclohexanone than a natural sniff. However, for butanol, Laing (1983) observed no differences in perceived magnitude when a strong or natural sniff was used. One problem with comparing these studies is the apparatus used. Although Laing (1983) had participants sniff at a port, Teghtsoonian and colleagues, in their various studies, had participants place one nostril over a glass tube. It is conceivable that these changes in equipment might account for the failure to obtain a constancy effect with cyclohexanone. A further source of variance should also be noted—that of the particular chemical used in each study. Mozell, Kent, and Murphy (1991) demonstrated that flow rate of an odorant over a frog's olfactory epithelium only increased response magnitude when the odorant was rapidly absorbed by the olfactory mucosa. Less well absorbed odorants did not demonstrate this effect. The implication of this finding, if it also applies to humans, is that constancy may be an illusory effect for some odorants, because the concentration at the receptor may not shift dramatically with a more vigorous sniff.

Both Laing's (1983) study and more recent reports (Johnson, Mainland, and Sobel 2003) have confirmed that sniffing patterns differ as an odorant's concentration is increased. In a detailed investigation Johnson, Mainland, and Sobel (2003) found that after approximately 200 ms, sniffing rates diverged depending on the concentration of the inspired odorant. High concentrations reduced sniffing rate. The neural basis of this effect may be mediated by the cerebellum, in that functional magnetic resonance imaging reveals that this structure is sensitive to changes in odorant concentration and that this information may then be used to modulate the vigor with which the odorant is sniffed (Sobel et al. 1998). This may reflect a further innate adaptation to maintain qualitative constancy by reducing the amount of odorant reaching the olfactory mucosa. Nonetheless, the findings of Mozell, Kent, and Murphy (1991), noted above, need to be considered here too.

In sum, the evidence here suggests a dynamic system, in which sniff vigor may modulate the perception of intensity, but in which odorant concentration also modulates sniff vigor. One paradox of this description concerns changes in odor quality resulting from more concentrated odorants binding to more receptors. Although sniff vigor constancy may be fine for modulating the perception of intensity, it is hard to see how it could compensate for changes in odor quality that may ensue with higher concentrations. As far as we are aware this has not been investigated, although, as we describe below, quality clearly can change with concentration.

Variation in Odor Quality with Concentration

Nearly a century of casual observation suggests that some odorants smell qualitatively different at high concentrations, as detailed in table 4.1. One might expect this type of finding, given that higher concentrations of any odorant are likely to result in progressively more widespread binding to different receptor types (Sicard 1990; Malnic et al. 1999). Empirical evidence favoring changes in quality with concentration have emerged from several recent human studies. In addition, animal work, although of a different form, because quality cannot be assessed directly, is also suggestive of the same conclusion. These findings are reviewed in this section as are their implications for olfactory perception.

The first formal investigation of whether quality changes with concentration was reported by Gross-Isseroff and Lancet (1988). They examined whether human participants judged the same odorant to be qualitatively the same or different, from the same odorant presented at a different concentration (i.e., A vs. A'—same or different?). Using three concentration steps, at one log step intervals, and six odorants (benzaldehyde, isoamyl acetate, diphenyl methane, diphenyl ester, citral, and eugenol), they found that participants were generally poorer at judging as the same the same odorant at

Table 4.1 Changes in Odor Quality with Concentration

Odorant	Quality at low concentration	Quality at high concentration
Macrocyclic ketones[1]	Musky	Cedarwood
Tetrahydrothiophene[2]	Coffee	Sulfurous
p-Meth-1-ene-8-thiol[3]	Grapefruit	Sulfurous
Civet[2]	Musky	Fecal
Ambergris[2]	Oceanic	Rotting
Heptanol[2]	Fragrant	Oily
Camphor[2]	Urine	Aromatic
Phosgene[4]	Hay	Silage
Furfuryl mercaptan[5]	Coffee	Fetid
Diphenyl methane[5]	Geranium	Orange
Diphenyl ether[5]	Geranium	Musty
Ethylamine[5]	Fishy	Ammonia
Indole[5]	Jasmine	Fecal
Methyl heptinoate[5]	Violets	Foul
Geosmin[5]	Earthy	Aromatic

[1]Amoore 1977
[2]Vroon 1997
[3]Rossiter 1996
[4]National Research Council 1943
[5]Gross-Isseroff and Lancet 1988, table 1

different concentrations than they were at control judgments. That is, the results indicated that changes in concentration did indeed affect odor quality, but there were considerable individual differences.

In a second experiment, they obtained further evidence consistent with changes in odor quality as concentration increases. They used four odorants this time; the one most consistently found to alter in quality with concentration in experiment 1 (benzaldehyde), the one least likely to change (citral), and two other odorants for which there was no a priori reason to expect any change (geraniol and L-carvone). As in experiment 1 same/different judgments were made, with an additional task this time of similarity ratings using a continuous scale. They found a generally strong relationship between the two forms of rating, but interestingly enough, each of the four odorants exhibited qualitative changes over the three concentration levels used. Participants were, as in experiment 1, poorer at judging say citral to be the same, when the two to-be-compared stimuli differed in concentration.

Although this study does seem to indicate that quality may change with concentration, one must be concerned that intensive cues may have been inadvertently used in making the same/different or similarity judgment. However, such a criticism cannot be made of a more recent human study reported by Laing et al. (2003), which obtained findings largely supportive of Gross-Isseroff and Lancet (1988). In this case, using a much larger sample size ($n = 37$ compared with $n = 8$ and 13), they started by obtaining thresholds for five compounds; 1-heptanal, methyl heptanoate, 2-octanone, 1-heptanoic acid, and 1-heptanol. Then 1-, 3-, 9-, 27-, 81-, 243-, and 729-fold concentration increases above individual thresholds were prepared for each participant, and each of these was rated using the 146-item Dravnieks quality rating scale (Dravnieks et al. 1986). The findings were very clear. For all of the odorants, changes in reported quality, on the average, altered as the concentration increased. For 1-heptanal, the quality reported went from a fragrant odor at low concentration to an oily smelling odor at high concentration; for methyl hepatanoate, it went from vanilla to pineapple; for 2-octanone, from woody to fruity; for 1-heptanoic acid, from floral to paint/chemical; and for 1-heptanol, from cucumber to citrus. Notice how these findings, although based on subjective ratings, cannot be accounted for by confusion between intensity and quality, unlike the earlier reported studies. More important though is how the two approaches complement each other in demonstrating that both subjective and objective techniques suggest the same outcome; quality changes with concentration.

Animal data also suggest a similar conclusion. Couread et al. (2004) found that rabbit pups would only respond to the mammary pheromone when it was presented within a fairly limited concentration range. That is, when weaker or stronger than the operative range, no behavioral response (rearing, oral seeking) was observed. This is consistent with the notion that concentration actively recruits more receptors, thus changing the quality of the receptor output. Moreover, it again points to the inadequacy of strictly labeled line approaches even for a class of stimuli (pheromones) that might be considered the archetype for such a model.

More direct evidence for recruitment of receptors as concentration increases has been obtained in two further studies. Fried, Fusst, and Korsching (2002) found that for a range of aldehydes in mice, at low concentrations, approximately ten to twenty glomeruli were activated and not much overlap occurs between aldehydes. At higher concentrations, about eighty glomeruli were activated, with much more overlap between aldehydes. Finally, Johnson and Leon (2000) found that for two odors that change in quality with concentration in humans (pentanal and 2-hexanone), the proportion of active glomeruli in the rat olfactory bulb changed significantly as concentration increased. For odorants for which there is relatively little qualitative change in humans with increased concentration, namely pentanoic acid, methyl pentanoate, and pentanol, the pattern of glomerular activity in rats remained largely unchanged as concentration increased. In sum, these findings suggest that primary reliance for odor quality is placed on the overall pattern of glomerular activity. When this changes, as it may often do with increases in concentration, so does the quality that the participant perceives. This implies that odor quality perception appears to depend on the pattern of activity rather than on the activation of specific receptors or glomeruli. This is consistent with the general theme of our argument, that pattern (i.e., a constellation of features forming an object) recognition underpins odor perception. Object recognition allows for some generalization, but with large intensity-induced changes in input pattern, perceptual experience changes.

CONCLUSION

Three conclusions emerge from this survey of the effects of stimulus intensity on olfactory perception. First, a role for object recognition is suggested by two findings. The observation that changes in odor quality, quite dramatic in some cases, can occur as odorant concentration increases, is suggestive of a reliance upon the pattern of stimulation. As increases in concentration are known to affect the selectivity of binding to receptors, this suggests that it is the overall pattern that is used to generate the olfactory percept, not a labeled line between specific receptors and specific qualities. If labeled lines were the norm, then although other "lines" would come on stream as concentration increased, the system should be able to discount these relative to the most powerful signal from the specific labeled line—as with taste.

The second conclusion is that experience can play a significant role in processes which appear to be wholly under biological control. Experience can alter such basic psychophysical parameters as threshold and suprathreshold judgments of intensity, although whether the former relies primarily upon up-regulation of gene expression and/or changes in CNS processing is currently unknown.

The third conclusion is that psychological factors play an important role even in processes that might be presumed to operate at a very low (receptor) level. Sniff vigor constancy affords one example, although there are some interesting concerns about how this relates to the finding of shifts in odor quality with concentration. If changes in concentration lead to changes in odor quality, what effect does vigorous or weak sniffing have on odor quality? As far as we are aware this has not been investigated. In sum, stimulus intensity factors do not impose any severe constraints upon a perceptual system that may operate via object recognition. Rather the data reviewed here suggest that object recognition and dependence upon experience are the norm, even in situations in which we might expect considerable biological constraints.

5

Odor Quality Discrimination in Nonhuman Animals

The Function of Animal Olfaction

As discussed briefly in chapter 1, olfaction, and chemosensation in general, plays a major role in day-to-day survival of all animals. Chemical cues are used for finding food, identifying mates and recognizing kin, avoiding predators, and, for many species, aid in territorial marking, homing, and navigation. Examples include the use of odorants by birds and bees to locate foraging sites or the use of chemical cues by invertebrate parasites to identify appropriate hosts. Mice and many invertebrates use chemical cues to recognize genetically related kin to avoid inbreeding or to foster communal care. Many animals can determine reproductive state or even the identity of specific individuals through olfactory cues. The presence of predators can be detected and, through alarm pheromones signaled to others with the chemical senses. Finally, home territories or nesting sites can be determined with chemical stimuli.*

The breadth of uses of chemical cues by animals suggests that the chem-

---

* An outstanding review of the use of odors and pheromones in animal behavior can be found in Wyatt, T. D. (2003) *Pheromones and Animal Behavior.* This book serves as the source of many of the examples cited here.

ical senses are very effective information processors. Informational content can be derived either from highly evolved stimulus-receptor relationships (for example, pheromonal activation of specific chemical receptors and associated central circuits) or from complex stimulus mixtures that allow sufficient variation for different odor mixtures to acquire different meanings (for example, complex odor mixtures that vary sufficiently to be used for odor recognition of individuals). The first process, in which information contained in simple molecules or mixtures is extracted with highly selective receptors and dedicated central circuits, can allow a high-speed, highly reliable translation of a chemical stimulus into a specific behavioral response. However, the specificity of simple signals limits the information content, and dedicated circuits and behaviors limit behavioral flexibility and adaptability if stimulus quality or contingencies change.

The second process, in which information is carried by complex mixtures (see fig. 5.1), enhances potential informational content by increasing the potential variation between different signals, but it raises new problems by requiring central processing that allows assignment of meaning to these complex signals. Furthermore, complex signals may be less reliable than single molecules or simple mixtures, in that individual components may occasionally be missing, background odors may interfere or overlap, or component concentration ratios may be variable. Thus, processing complex chemical signals may require a more synthetic, dynamic approach in which unitary percepts are created from the complex mixture, producing reliable perception in the face of signal degradation or background interference. The majority of perceptual odor objects we perceive in the world — coffee, Bordeaux, perfume, and chocolate — are complex mixtures of often hundreds of molecular components, yet they are perceived as single odor objects or olfactory gestalts. A major thesis of this chapter and this book is how complex mixtures come to acquire a unity of perceptual quality and behavioral meaning.

SENSORY ECOLOGY

Communication and information exchange in all sensory modalities is affected by the environment through which the signal is transmitted and by the characteristics of the receiver (Dusenbery 1992). These effects include signal degradation or interference over time and distance, and signal versus background noise interference. For example, in terrestrial ecosystems, chemical stimuli can vary widely in volatility, which affects the duration that a signal

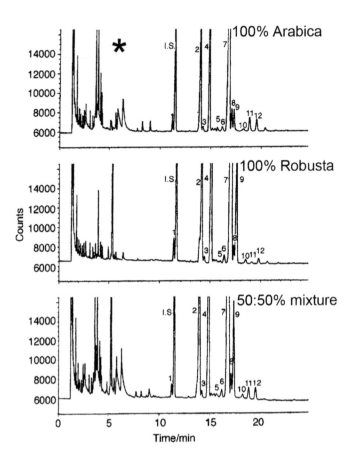

Fig. 5.1. Gas chromatograph analyses to three coffees discriminable by humans based on their odorants. Note that whereas all three stimuli are complex mixtures with many overlapping components, each unique combination produces a perception distinct from the others, similar to the pheromonal mixtures in insects described here. Adapted from Valdenebro et al. (1999).

lasts and the speed with which it is transmitted—a fact that can be taken advantage of in chemical communication. Thus, pheromones used to mark territories tend to be composed of large, less volatile molecules that allow marking specific locations for a relatively long time. In contrast, alarm pheromones tend to be composed of small, more volatile molecules, allowing rapid diffusion over short distances and quick dispersion, thus providing a fast, brief signal.

However, the volatile nature of the stimulus also raises problems for detection, discrimination, identification, and source localization. Chemical stimuli have an active space around the source, generally defined as the spatial region within which the stimulus is above the detection threshold of the receiver. Unfortunately, as described elsewhere, the perceptual quality of some odorants changes with concentration (Doty et al. 1975). Thus, this raises the need for mechanisms, at least in some cases, of perceptual constancy.

On the other hand, chemical stimuli such as released pheromones can often disperse as patches, rather than as a uniform gradient away from the source. These patches are distributed by air or water flow around the source. This results in a complex topography of relatively higher stimulus concentration downstream of the source intermixed with regions of no or very little stimulus. This may help solve problems of perceptual constancy but raises new issues for source localization.

In addition to varying in concentration, chemical stimuli and chemical mixtures can vary in quality over time. For example, components of mixtures may be differentially degraded or dispersed with time. Thus, an odorant source may produce an apparently variable or incomplete chemical signature. Under these conditions, mechanisms for stimulus completion of a partially degraded signal could be beneficial for discriminating and identifying odors in the real world.

Finally, background odors may be (most likely are) present that could interact or interfere with processing of the target stimulus. Given that the current views of olfactory coding at the receptor sheet involve submolecular feature analysis, on any given inhalation or sweep of chemosensory antenna, a jumble of receptors will be activated—some by the features of the target stimulus and some by background odorants. The olfactory system thus must somehow sort out how to group the features—which features belong with the target (itself potentially a complex mixture) and which should be treated as background. The likelihood that there will be overlap in features between the target and background adds to the issue of stimulus completion noted earlier.

Table 5.1 Factors Affecting Odorant Transmission

Factor	Effects		
Stimulus factor			
Size/volatility	Diffusion rate	Stimulus persistence	
Initial concentration	Diffusion distance	Stimulus persistence	
Pulsed release	Diffusion distance		
Environmental factor			
Temperature	Diffusion rate	Stimulus persistence	
Media movement	Diffusion rate	Diffusion distance	Stimulus persistence

As discussed in chapter 2, none of the issues raised here are unique to olfaction and odor stimuli. Dealing with degraded images, perceptual constancy, perceptual gestalts, and perceptual grouping are all necessary to varying degrees in other sensory systems. What is perhaps unique about the olfactory system, however, especially in comparison with vision or somatosensation is the lack of an external spatial dimension within the stimulus to help in solving these problems. For example, in vision, perceptual grouping of features can be facilitated if all the features move coherently across the visual field. This spatial component of the stimulus is missing in olfaction.

The next section divides olfactory information processing by nonhuman animals into four main task components (detection, discrimination, intensity judgment, and assignment of meaning) and describes examples of how these different behavioral expressions of information processing are dealt with, or ignored, by various species for a variety of tasks. The subsequent sections describe how olfactory systems have evolved to deal with these different issues in different contexts and the consequences for understanding odor perception.

What Are the Necessary Conditions for Use of Odor Information?

To use chemical cues for the functions and under the conditions described above, both vertebrates and invertebrates must have olfactory systems that can deal with at least four specific issues. These issues include (1) detection of the stimulus; (2) discrimination of the stimulus from other, potentially very similar stimuli; (3) some way to deal with and perhaps determine relative stimulus intensity; and (4) some method of assigning meaning to the stimulus. Examples of each of these are discussed below.

Stimulus Detection

First, the stimulus must be detected, in general, against other, background stimuli that may or may not have biological significance. This background may be correlated with the stimulus of interest (i.e., form a coexistent mixture) or may be uncorrelated with the stimulus (truly background). In rodents, for example, the proximity of predators can be determined by chemical cues, with cat odor or fox odor (among others) producing characteristic stress and unconditioned fear reactions in rats. Of course, it would be maladaptive to be continually stressed and fearful, thus correct identification of the potential predator odor and discrimination of it from other odors is important. In rats and mice, trimethylthiazoline (TMT), a single volatile component of fox feces can be sufficient to induce fear reactions. Under either laboratory or natural conditions, this odorant may occur against a background of dozens or hundreds of different odorants, yet the animals are capable of detecting that signal and acting appropriately.

Salmon migration similarly appears to depend in part on detection of specific chemical cues that are set against complex backgrounds. Salmon hatch in small freshwater streams where they stay for several months until the developmental physiological transition (parr-smolt transition) that allows them to adjust to subsequent life in the salt water of the open ocean. At this parr-smolt transition, hormonal changes occur that allow olfactory imprinting on the chemical signature of their natal stream (Nevitt and Dittman 1998). Several years later when the adult salmon return from the sea to freshwater to reproduce, they return to their natal stream, based on this olfactory memory. In a wonderful series of experiments, captive bred young Coho salmon were imprinted onto the artificial odorant phenyl ethyl alcohol before being released into Lake Michigan (Scholz et al. 1976). When the adult salmon were to return to their home streams for mating, low levels of phenyl ethyl alcohol were added to one freshwater stream and counts were made of how many released salmon entered the target, experimentally odor-labeled stream, and how many entered nearby, unlabeled control streams. As predicted, salmon artificially imprinted on phenyl ethyl alcohol in the lab entered the phenyl ethyl alcohol–labeled stream—a stream they had never been in before. Obviously, this stream, and the many control streams, contained a variety of chemicals that may or may not have been detected by the olfactory systems of these fish, but the critical imprinted odor was detected against this background.

Similar results are seen in invertebrate pheromone detection, where a specific odorant (or odorant mixture) is detected and acted on, in general, despite the presence of many co-occurring background odorants. For example, undecane is an alarm pheromone in carpenter ants (*Camponotus*) that, when detected, induces arousal and aggression in ants receiving the pheromone. As in the salmon-imprinting example, this pheromone is detected and effective in a variety of background odor contexts.

Finally, in perhaps one of the penultimate examples of need for and success at detection of a specific chemical signal against an odorous background, golden hamsters (*Mesocricetus auratus*) use chemosensory cues to recognize other kin and nonkin individuals (Todrank, Wysocki, and Beauchamp 1999). Golden hamsters have several scent sources, including flank scent glands, ear scent, urine, and female vaginal secretions. Each of these scents varies from one another in the same individual and, as a whole, varies between individuals. In cross-habituation tests of discrimination, Johnston and colleagues have found that a familiar individual can be recognized by any one of these individual scents (Johnston and Bullock 2000). Both male and female hamsters scent mark on surfaces in their environment and will competitively overmark (or countermark) scents of other individuals. Remarkably, hamsters can determine which odor cue, and thus, which corresponding marking individual is most recent (i.e., marked on top; Cohen, Johnston, and Kwon 2001). Similar findings have been reported in mice (Rich and Hurst 1999). Although determining which mark is on top involves some spatial analysis of the marks, it clearly also requires discriminating a complex odor cue against a closely related complex background. Johnston has interpreted this perceptual ability as olfactory scene analysis involving recognition of multiple odor objects from each other and from background.

Together, these examples demonstrate that stimuli ranging from specific simple molecules to complex mixtures can be detected and influence ongoing behavior. This detection is made more difficult by the presence of background stimuli that may or may not be similar to the target. Nonetheless, animals across the species spectrum and across development appear capable of solving this detection problem.

Stimulus Quality Discrimination

The second problem that must be dealt with for effective use of chemical cues, and clearly is involved in the golden hamster example above, is that the stimulus must be discriminated from other, perhaps very similar stimuli. In laboratory situations, many animals are capable of making behavioral discriminations of pure odorant molecules containing hydrocarbon chains varying by a single methyl group (one carbon and its associated hydrogen ions). Similarly, behavioral discriminations of chemical enantiomers can be made —that is, discrimination between two identical volatile molecules varying only in their structural conformation. Thus, for example, rats can be trained to behaviorally discriminate between (+)-limonene and (−)-limonene or between (+)-carvone and (−)-carvone, although these molecules are structurally incredibly similar. Discrimination of the carvone enantiomers can occur in naïve mice, while discrimination of the limonene enantiomers requires training. This distinction is discussed in more detail in a later chapter.

Another aspect of stimulus quality discrimination is the often unique perceptual properties evoked by stimulus mixtures compared with their individual components. Simple mixtures can be perceived in either an elemental form or in a configural form. Elemental or analytical mixture perception implies that the individual components comprising the mixture (or a subset) provide recognizable features to the mixture percept. Thus, when stimulated with a binary mixture of banana and orange odors, the subject may be able to identify that both banana and orange are present. In configural or synthetic mixture perception, on the other hand, the mixture is perceived as unique from the components and, in fact, the individual components may be impossible to identify.

Mixtures containing more than three components appear, in general, to be perceived configurally. However, simple binary mixtures can be perceived either configurally or elementally, depending in part on the nature of the components. In rats, binary mixtures of molecularly similar odorants appear to be perceived configurally, thus producing a mixture percept distinct from the components. This allows easy discrimination between the mixture and its components but of course limits analysis and identification of those components. In contrast, binary mixtures of molecularly dissimilar odorants appear to be processed analytically, producing reduced discrimination of the mixture from its components and enhanced ability to recognize or identify those components (Kay, Lowry, and Jacobs 2003; Wiltrout, Dogra, and Linster 2003).

In naturalistic situations, similar remarkable feats of olfactory acuity have been demonstrated and, in fact, appear critical for adaptive behavior. For example, many invertebrate parasites identify their hosts based on chemosensory cues. These parasite-host relationships have often evolved to be highly selective, with a particular parasitic species depending entirely on a single host species, discriminated from other potential hosts through chemosensory cues. For example, several species of Hymenoptera that are parasites of aphids are attracted to their hosts by the female aphid sexual pheromone nepetalactone. *Aphidius ervi*, a parasitoid of the pea aphid is attracted to only one enantiomer of nepetalactone (7S), the presence of the other enantiomer (7R) is not attractive, and a mixture of both enantiomers (1:1 mixture of 7S:7R) reduces the attractiveness overall (Glinwood, Du, and Powell 1999). Thus, *A. ervi* are capable of making this fine discrimination between enantiomers to locate their host.

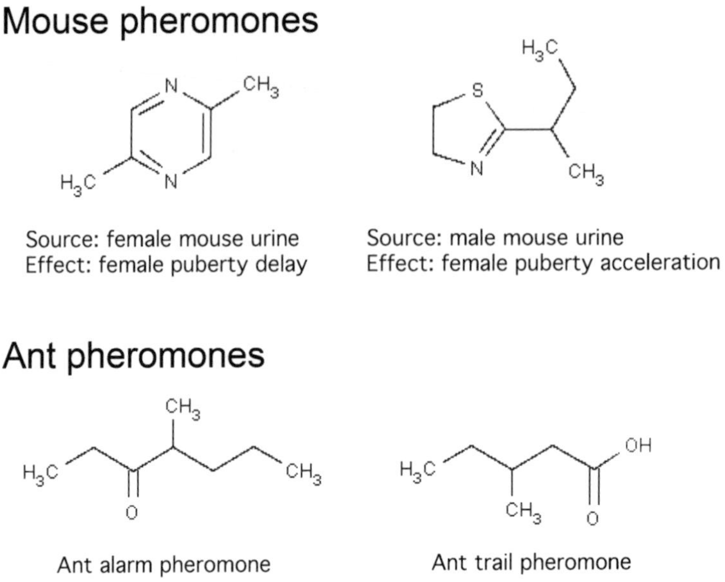

Mouse pheromones

Source: female mouse urine
Effect: female puberty delay

Source: male mouse urine
Effect: female puberty acceleration

Ant pheromones

Ant alarm pheromone

Ant trail pheromone

Fig. 5.2. Structures of chemical signals that produce very specific behavioral or physiological responses in insects and mammals. The mammalian examples here primarily function via the vomeronasal system; however, increasing evidence suggests a role for the main olfactory system in sensation and perception of urine volatile in mammals. (A) Examples of mouse pheromones. (B) Examples of ant pheromones.

Similar highly selective relationships exist between some invertebrates and plants. In egg-laying female moths, volatiles from leafy plants are signals for appropriate ovipositing sites. Identification of the appropriate plants for ovipositing is critical for larval survival and to reduce competition between phytophagous insects. In *Manduca sexta*, females are attracted to tobacco plants or similar plant species such as tomatoes (family: Solanaceae), where they lay their eggs, and are not attracted to other plants. *Manduca* larva are capable of dealing with the insect-killing toxins these plants contain (such as nicotine), and feed until pupation. Leaves and flowers of these plants release scores of different volatile compounds, only a few of which guide moth orientation.

As another example, in many altricial mammals newborns must recognize their mother to acquire life-sustaining care, and the mothers must recognize their offspring to selectively devote their resources to the survival of their genetic relatives. As noted above, a major component of this mother-infant recognition is mediated by olfactory cues. In sheep, the ewe only allows her own lamb to suckle, which she identifies through chemosensory cues. Lambs may all smell alike to humans, but very subtle differences (currently largely undefined) are sufficient to allow a lactating ewe to identify her lamb as distinct from other lambs of the same age. Conversely, neonatal rodents recognize and orient to their mother based on olfactory cues. In rats, these odor cues are learned through pre- and postnatal experience and evoke behavioral activation and nipple attachment behaviors critical for survival. Such behaviors are not evoked by odors of unfamiliar postpartum, lactating females. Similar odor-evoked behaviors are observed in human infants within hours after birth in response to maternal odors, but not unfamiliar postpartum female odors.

A variety of other kin recognition and mate recognition systems based on odor discrimination exist in mammals. Perhaps the best described relate to recognition of odorants controlled via the major histocompatibility complex (MHC) genes. MHC genes are a highly diverse set of alleles providing important individual-specific identity cues for immune system recognition of self. Volatile MHC-related products are released through the urine and can serve as phenotypic markers of individual identity. Mice use MHC-derived odors to identify relatives and nonrelatives in situations where joining with genetic relatives (e.g., communal care of young) and avoiding genetic relatives (e.g., mate selection and outbreeding) are advantageous. The MHC-derived odors are sufficiently different that even rats can be trained to behaviorally discriminate between different strains of mice.

The examples thus far clearly outline both the ability and the need for animals to detect and discriminate between odorants or odorant mixtures, often molecularly very similar and generally against some background. In addition to these very specific examples of biologically significant and/or biologically determined odorants (e.g., pheromones), often tied to specific behavioral tasks (e.g., mating or host recognition), many animals must also be capable of detecting and discriminating a wide range of odors in more varied contexts. For example, animals that pollinate or feed off of a wide range of flower species, such as bees, must be able to discriminate a large set of different floral odors. Floral odors are one of several cues used by bees to locate foraging sites. Similarly, frugivorous primates and mammals must make discriminations between edible and nonedible or ripe and nonripe fruits. In fact, the olfactory systems of frugivorous nocturnal mammals tend to constitute a larger proportion of the central nervous system than in diurnal frugivorous mammals, because the nocturnal mammals rely on olfaction to make these feeding decisions, whereas the diurnal mammals rely more heavily on visual cues

(Barton et al. 1995). Humans, for example, are more likely to be attracted to a bright red apple, regardless of its scent, whereas a nocturnal lemur will be guided by the scent, a fact not overlooked by the local grocer. In addition, in some of the preceding examples, we have raised the issue that some odorants only become important, and thus worthy of discrimination, after experience (e.g., odors involved in mother-infant interactions). Thus, discrimination (and perhaps detection) can be influenced by experience. This is an issue we will explore in greater detail below.

The other issue raised by these examples, besides the importance of discriminating odors from each other and as different from background, is that frequently, perhaps normally, odors of biological significance for animals are complex mixtures of many components, not monomolecular odorants. The critical difference in odors of two genetically unrelated mice stems from the unique combination of volatile components in the mice's urine. The attractiveness for a male *Manduca* moth of sex pheromone released by a female *Manduca* is produced by the specific combination of components in the pheromone mixture. It is the mixture that is important; presentation of one component alone does not produce partial attraction, it produces no response at all in the male. In fact, as we have argued elsewhere, olfactory systems excel at synthesizing multicomponent, complex mixtures into unique olfactory perceptual objects. While the devil may be in the details, much of olfactory perception, and thus odor-guided behavior in nonhuman animals, is in the synthetic whole.

Stimulus Intensity Determination

The third problem that must be dealt with for an animal to function effectively with chemosensory stimuli is variation in stimulus intensity. Odorants diffuse and are carried by currents in air or water away from their sources, and thus stimulus concentrations can vary substantially over very short distances. In some situations or for some tasks, stimulus intensity provides critical information, for example, in tracking or homing situations requiring movement within a concentration gradient (although see below). Identifying intensity (or relative intensity) of components within mixtures may also be important where a specific blend or ratio of components is the meaningful stimulus, rather than a single molecule. On the other hand, variation in stimulus intensity can sometimes interfere with stimulus quality identification and discrimination — the issue of perceptual constancy noted above. As discussed elsewhere, for example, olfactory receptor neurons become much more broadly tuned (less selective) as stimulus intensity increases, and in humans, the perceptual quality of some odorants can change dramatically as absolute concentration changes. Similarly, in laboratory rats and honeybees, discrimination of odorants or odorant mixtures is stimulus intensity dependent.

Localizing an odor source can be performed in at least two primary ways. First, chemotaxis involves movement in a chemical concentration gradient either toward higher concentrations of an attractant or toward lower concentrations of a repellent. True chemotaxis thus requires knowledge of relative stimulus intensity and some way of comparing intensity while moving in the gradient. Comparison of changes in intensity as one moves through a concentration gradient (am I moving up or down the gradient?) is performed in two manners. Some species use simultaneous comparisons of intensity at two different sites on the body, tropotaxis. This process of course requires a body size sufficiently large relative to the concentration gradient to detect differences between the two sensors. As described below, this requirement may help account for the size of a snake's forked tongue. Other species use se-

Table 5.2   Mechanisms of Stimulus Localization

Movement within chemical gradients can be guided by several distinct mechanisms.

Kinesis—Movement that depends on stimulus intensity, but individual turns and movements are not directly related to the stimulus concentration
  Orthokinesis—movement speed varies with stimulus concentration
  Klinokinesis—rate of turning or rotational movement varies stimulus concentration

Taxis—Movement that depends on stimulus intensity, where movement across a concentration gradient is mediated by turns related to the stimulus concentration
  Klinotaxis—Sequential sampling is used to establish the concentration gradient and direct movement
  Tropotaxis—Simultaneous sampling from two points on the body is used to establish the concentration gradient and direct movement
  Teleotaxis—An extreme form of tropotaxis not known to occur in chemical senses involving simultaneous sampling from many points on the body and subsequent orientation in relation to this patterned array

Each of these specific mechanisms can be further defined through prefixes such as *chemo-* (relative to chemical stimuli), *anemo-* (relative to wind currents), and *rheo-* (relative to water currents). Thus, rheotropotaxis would involve simultaneous sampling of water currents on two points of the body and movement relative to that current. Such behavior could be initiated by detection of a chemical attractant, and movement through the current could lead the animal to its source.

List adapted from Wyatt (2003) and Dusenbery (1992).

quential comparison of stimulus intensity, where intensity is determined at one site, the sensor moved and the intensity sampled again. Temporal, sequential sampling of this type can be performed by very small organisms and in very shallow concentration gradients, but it requires a memory of initial intensity to allow comparison with later samples.

As opposed to true chemotaxis, the second method of localizing an odor source and movement within a concentration gradient actually does not involve determination of concentration. Rather, the attractive odorant leads to anemotaxis, movement directed relative to the wind.* Thus, in M. *Sexta* males, the attractive female sex pheromone causes the animal to begin a stereotyped upwind movement, with the expectation that the source will be upwind of the current position. In this case, determination of precise or even relative concentration is not required, only detection and discrimination. This is actually an adaptive way to solve the problem of locating an odor source because odor plumes are generally not strict concentration gradients diminishing from the source, but rather patchy pockets of potentially equal concentration. Each time the male hits one of these pockets, upwind flight is maintained until the source is reached. In fact, placing a *Manduca* male in an artificially prepared constant odor gradient, without patches of odorous and nonodorous air, retards upwind flight and source localization. Hence the patchy structure of the odor plume—high concentration wisps surrounded by clean air—is part of the signal leading to anemotaxis.

Decapod crustaceans such as lobsters, crayfish, and crabs use odor plumes to locate food and conspecifics, and orientation to chemical sources has been studied in detail in these species (Grasso and Basil 2002). These crustaceans have chemosensory hairs located on both their antennules and legs, allowing sampling at various heights from the floor. In addition, something akin to active sampling may occur when the lobster or crayfish flick their antennas. Flicking is a stereotyped whiplike movement of the antenna that appears to enhance stimulus access to the receptor sites on the antennule chemosensory hairs, perhaps similar to active sniffing in mammals.

Several different mechanisms have been proposed for decapod orientation in odor plumes and localization of odorant sources. The first is similar to that described for the *Manduca* above and essentially involves anemo- or rheo-

---

* Anemotaxis involves movement relative to wind currents and rheotaxis involves movement relative to water currents. Current direction can be determined both visually and mechanically, the latter by sensing movement of hair cells on the body surface. In many invertebrates, the antennules, which contain the chemoreceptors, also are important mechanosensory organs.

taxis evoked by odor detection. Thus, stimulus concentration determination per se is not required for movement toward the source, only stimulus detection. However, in American and spiny lobsters, two intact antennules are required for orientation and plume tracking. This has led to two alternate hypotheses that involve simultaneous comparisons of stimulus concentration on bilateral chemosensors. One of these mechanisms is tropotaxis, where the chemical concentration gradient is determined by direct comparison of stimulus concentration on two spatial distant receptors (e.g., the two antennules or two chemosensor bearing legs). If the concentration of an attractant is relatively higher on one side than the other, then the animal turns in that direction. Another hypothesized mechanism is plume edge tracking, where one chemosensor is maintained within the odor plume and the other is kept outside the plume. This is similar to tropotaxis, but would not require comparisons (or knowledge) of relative stimulus concentration; simple detection could account for this behavior. Other mechanisms involving more complex analysis of currents and eddies with mechanoreceptors combined in rheotaxis have been posited. These also do not require determination of stimulus concentration, however; instead, they take advantage of the fact that odor plumes tend to be composed of highly concentrated wisps of odorant surrounded by odor-free zones and movement of the media in which the odorant is contained. Under these conditions, whether aquatic or terrestrial, determination of absolute or relative concentration is generally believed not required for approach to an odor source.

However, some organisms do use information about relative concentration more directly. For example, *Escherichia coli* bacteria move in liquid chemical concentration gradients toward higher concentrations of attractants like sugars and amino acids. As long as the concentration of the attractant continues to increase as the bacterium moves in its environment, net movement will continue up the concentration gradient. If the attractant concentration does not increase as the bacterium moves, net movement toward the attractant ceases. Thus, relative stimulus intensity is critical for directing the movement of *E. coli* toward nutrients. Similarly, very slowly moving starfish appear to determine average relative stimulus concentration in one location and compare it to average stimulus concentration in the previous location. Because of their very slow movement, this averaging allows temporal filtering of high and low intensity patches (Grasso and Basil 2002).

Two vertebrate examples of animals studied in some detail that may use relative stimulus intensity in localizing or tracking odor sources are snakes and dogs. Venomous snakes use at least two different predation techniques.

Some species hold their prey after injecting their venom until the prey is sufficiently incapacitated to be consumed. Other species, such as rattlesnakes, inject their prey and then release them while the venom takes effect. This latter strategy requires that they then be able to track the prey after it has run off. They can do this tracking through chemosensory cues that they acquire through their vomeronasal system (also known as Jacobson's organ) and tongue flicking. Tongue flicking can be used for sampling chemical cues either by touching the tongue fork tips directly on a substrate (like the ground or the prey individual) or waving the fork tips in the air. Chemical stimuli adhering to the tongue fork tips are then applied to the bilateral vomeronasal organs in the roof of the mouth as the tongue is retracted. Snakes that track their prey have significantly longer forks than snakes that inject and hold their prey. It is hypothesized that longer forks allow greater tip separation and thus greater sensitivity necessary for tropotaxis in shallow concentration gradients as simultaneous comparisons are made between stimulus intensity at the two tips.

Dogs appear to track odors with sequential sampling and temporal comparisons of intensity. If a dog is brought perpendicularly toward an odor track left by a walking human, the dog samples several discrete neighboring footprints and compares stimulus intensity between them to determine the direction of travel. There is significant variation in the ability of dogs to perform this task, and training can enhance accuracy of direction determination. Nonetheless, the results imply an important role of stimulus intensity discrimination in performance of this task. To our knowledge, a direct test of odor tracking in dogs using scent marks with intentionally manipulated stimulus concentrations has not been performed.

These studies demonstrate that many animals use stimulus intensity discrimination to localize odorant sources, although for many species, odorant intensity discrimination is not required for source localization. The second major function of odorant concentration or relative concentration is as a source of information in recognition and response to odor mixtures. The effects of component concentration in mixtures of pheromones have been most thoroughly investigated, primarily in invertebrates. In many cases of invertebrate pheromonal mixtures, not only are specific components required to evoke a response in the receiver, but also these components must be in the correct concentration ratio. Modification of either the components or the ratio of component concentrations is sufficient to impair recognition and response to the pheromone.

For example, male oriental fruit moths (*Grapholitha molesta*) express a wing-fanning behavior in response to the female sex pheromone, which is a mixture of three components. The female pheromone consists of 91% Z8–12:Oac,* 6% E8–12:Oac, and 3% Z8–12:OH. Thus 91% of this pheromone by weight is the Z-isomer. However, in behavioral assays, Z8–12:OAc alone produces no significant male response; all three components, in the correct mixture ratio are required for evoking the male behavior. This suggests that the male has a synthetic percept of the mixture components, in the correct ratios, substantially different from the percept of the individual components alone. Both intensity ratio and component identity are important pieces of the synthetic whole.

Although relative odor intensity itself may be a critical part of the information contained in an odor stimulus while moving through concentration gradients and analyzing mixtures, absolute intensity can also influence stimulus quality discrimination. Although questions about effects of intensity on odor perceptual quality can most easily be addressed in humans where the subjects can directly classify their perceptual experience, the effects of stimulus intensity on odor discrimination per se can be examined in animals. In operant conditioning tasks, rats are very good at discriminating odorant concentration differences (Slotnick 1985). Thus for example, rats can learn to discriminate between a 0.1% concentration of proprionic acid and a 0.07% concentration of the same odor, though unable to discriminate between 0.1% and 0.08% concentrations. Intensity, therefore, has a perceptual quality capable of directing rat behavior.

Stimulus intensity also directly influences odor quality discrimination. Although odor discrimination at detection threshold concentrations is relatively poor as one might expect, odor discrimination of molecularly similar odorants is stable or somewhat enhanced as stimulus intensity increases in both rats and honeybees (Cleland and Narla 2003; Wright and Smith 2004). This suggests that once stimulus intensity rises above detection threshold, percep-

---

* Pheromones are frequently composed of mixtures of chemical isomers. The notation Z8–12:OAc and E8–12:OAc in this mixture refers to the fact that these two acetate compounds only differ in the specific geometric relationship between double bonds located between carbons 8–12 and the rest of the molecule. Thus, these two components are chemically highly similar, composed of identical units arranged slightly differently geometrically. However, isomers such as these can have distinctly different perceptual qualities. The isomers D-carvone and L-carvone, for example, essentially mirror images of each other, smell to humans like caraway and spearmint, respectively.

tual quality for many odorants (though not all) is largely constant over a wide range of stimulus intensity, similar to perceptual constancy of shape discrimination and identification in vision.

These examples demonstrate that animals can discriminate changes in stimulus intensity and that stimulus intensity can be used by animals for at least two functions—for source localization and as an informational feature of stimulus mixtures. In both cases, relative stimulus intensity, not absolute stimulus intensity, is important, similar to intensity coding in other sensory systems. Use of intensity in source localization is less common than might be thought, because many species instead use detection of the stimulus to direct upwind or up-current movement and do not appear to require determination of change in intensity to localize the source. Some species that do use stimulus intensity information for movement in a concentration gradient must include memory as a critical component of the process while making comparisons of temporally sequential samples of the current intensity.

Perhaps a more common role of relative intensity determination is the informational content it provides in stimulus mixtures. A 91:6 mixture of specific female moth pheromonal components will provoke upwind flight in the male of that species, whereas a 99:1 mixture of the same compounds produces no such behavioral response in the males.* It seems unlikely that the male moth is actively encoding that stimulus A is present at this concentration and stimulus B is present at that concentration and thus this is a female pheromone. Rather, as we argue below, the unique combination of odors A and B at the correct ratio evokes a single synthetic percept—"sexually receptive female moth"—that drives the behavior.

Finally, given that absolute stimulus intensity can influence odor quality discrimination, the issue of perceptual constancy across the intensity scale is raised. Animals such as some sea birds can identify food sources from hundreds of kilometers from the source, presumably at very low stimulus concentrations and continue to recognize that same stimulus throughout approach as concentration increases dramatically (Nevitt 2000). Olfactory perceptual constancy has not been well investigated in animals, but given the rather dramatic effects changes in stimulus intensity can have on both re-

---

* There are a variety of examples in other sensory systems where the tuning of the receiver's sensory system highly matches that of the sexual signal. For example, the tuning sensitivity of the auditory system of female Tungara frogs closely matches the frequency distribution of the male call. A similar match between signal characteristics and receiver sensitivity appears to occur in some invertebrate chemosensory sexual signaling systems.

ceptor and central processing (described below), it seems an important area of future investigation.

Stimulus Meaning

The fourth problem that must be dealt with for an animal to function effectively with chemosensory stimuli is assigning meaning to the stimulus. Once the odor is detected and discriminated from other odors, a behavioral response appropriate to that odor must be generated. Thus, meaning in this case refers to the fact that an appropriate, stimulus-related, behavioral response is evoked or made more probable by the odor. Examples might include aggregation pheromones that enhance locomotion toward the source,* or alarm pheromones that enhance locomotion away from the source. The meaning of each of these signals to the receiver is operationally defined by the behavior or potential behavior they evoke—one means move toward the source and one means move away from the source. In some animals, meaning may include hedonic or emotional value (e.g., preference or fear) perhaps with no overt behavioral response, though in most, this level of interpretation is not necessary to extract a concept of odor meaning.

INNATE MEANING

Odor meaning can either be innate or dependent on experience. Very often predator odors and pheromones have innate meaning shaped by evolution. For example, the attractants and repellents of *E. coli* described above have fixed meaning, although sensitivity to the stimuli may change over time through adaptation. Despite the fact that laboratory rats and mice have been isolated from carnivorous predators for many generations, odors of cats and foxes evoke behavioral and physiological responses consistent with fear. Cat odor evokes behavioral freezing and increases in the stress hormone corticosterone in laboratory rats despite having never encountered a cat in their lifetimes or in the lifetimes of their parents.

---

* Note that aggregation pheromones can function in at least two manners to enhance accumulation of individuals at the source of the pheromone. Some aggregation pheromones produce anemotaxis or chemotaxis to actively draw individuals nearer the source from some distance. Other aggregation pheromones function by arresting ongoing movement, such that individuals that happen to pass by the source are more likely to stop.

Similar to these predator odors, most pheromones* also have innate meaning. For example, the compound E-β-farnesene (a terpene) is an alarm pheromone in aphids and when detected enhances the probability of avoidance behavior in conspecifics. Note that pheromone-evoked behaviors are not totally reflexive. Aphid alarm pheromone is released in response to predation. Under most circumstances, the alarm pheromone will evoke avoidance in neighboring aphids or even cause the receiver to drop from the plant to the ground to avoid the predator (despite the potential high risks associated with exposure on the ground). However, the specific response to the alarm pheromone may vary depending on the quality of the plant being consumed or other local factors. Thus, while feeding on an especially rich food source, the avoidance response to the alarm pheromone may be reduced. Nonetheless, exposure to alarm pheromone always increases the probability of alerting, avoidance, or escape. Similar variations in response to alarm pheromones are seen in honeybee behavior, depending on the receiver's location relative to the hive. Again, the alarm pheromone is always interpreted as meaning a threat (operationally defined), although the specific appropriate response may vary depending on the context.

Similar to the alarm pheromones, sexual pheromones enhance the probability of reproductive behaviors and thus have a set meaning to the conspecific receiver. For example, male boar saliva contains 5-alpha-androstenone and 3-alpha-androstenol, which serve as sex pheromones to sows. Sows exposed to these compounds are more likely to exhibit a lordosis response (or standing behavior) in response to a mount by the boar, thus facilitating copulation, compared with females that do not detect the pheromone. In the same manner, pheromones in female hamster vaginal secretions increase the likelihood of male approach and mounting behaviors. These kinds of chemical stimuli, thus, have set meaning to the receivers.

---

* Pheromones can have many effects on the receiver and, in general, are classified as primer pheromones or releaser pheromones based on the nature of that effect. Primer pheromones produce long-term effects on the physiology of the receiver, such as delaying or accelerating onset of puberty. Releaser pheromones have more immediate effects on the receiver, such as evoking behavioral attraction or avoidance. Some pheromones can have both primer and releaser effects. This discussion focuses primarily on releaser effects. Finally, the notion of contextual effects mentioned in the text refers to the fact that some pheromones have multifunctional effects, depending on concentration, presence or absence of additional components, or physical location of the receiver (Ryan Insect Chemoreception, p. 100).

Under many circumstances, however, odor meaning is acquired through experience, often in a simple associative manner after temporal pairing of a novel odor with some biological, unconditioned stimulus. A fascinating example of this occurs in leaf cutter ants. Leaf cutter ants gather plant material to serve as a substrate for growing fungus on which they feed. These ants may select any of a variety of plants for this fungus gardening but continually monitor the health of the garden and may adjust their plant foraging accordingly. Ridley and colleagues (Ridley, Howse, and Jackson 1996) showed that ants provided with small bits of orange peel adulterated with a fungicide soon learned to reject orange peel once the fungus garden began to show poor health. It appears that the foraging ants learn the odor of their plant offerings (they are also more likely to acquire familiar scented plants than unfamiliar ones) and adjust their preference for particular species depending on the consequences for the fungus. The association of plant odor with fungal health appears to be derived by contact chemoreception with the gardening workers by the foraging workers. In a further example of acquired odor meaning, *Drosophila* in the laboratory can be taught to avoid odors associated with electric shock. *Drosophila* that received simultaneous exposure to a previously unfamiliar odor and electric shock changed their subsequent behavior toward that odor from random or no response to avoidance. Memory of the learned aversive odor lasts for many hours.

These examples not only demonstrate how odors can change or acquire their meaning based on experience, but also demonstrate that even invertebrate olfactory systems, often seen as largely reflexive pheromonal detectors, can include a more dynamic, flexible system component. Several additional examples of learned odor meaning are described below that include learned odor cues for kin and mate recognition and for learning food and foraging preferences. In each case, novel odors come to evoke behaviors consistent with a particular acquired meaning. As we argue elsewhere in this volume, we hypothesize that this meaning is not only associated with, or evoked by the odor, but actually becomes a component of the olfactory percept. These examples are meant to demonstrate the variety of channels used for associative change in odor meaning, including association between different chemosensory signals and association between odors and multimodal, nonchemosensory stimuli. They also demonstrate the variety of behavioral contexts in which learned changes occur. The inclusion of meaning to an odor percept is not an unusual event, but rather appears across the animal kingdom and thus, we hypothesize is an evolutionarily important component of olfactory processing.

Kinship and Mate Odors.   Knowledge of the genetic relatedness of other individuals to oneself is used by many species to direct energy and resources toward genetic relatives and reduce inbreeding. Although the goal of this individual discrimination is to determine genotypic relatedness, the animal must rely on phenotypic variation to provide the discriminatory cues, given that quantitative DNA analysis is not an option for most animals. For a cue to provide sufficient information to allow discrimination between individuals of the same species (i.e., signature cues), there must be substantial variability between individuals (often accomplished through complexity of the signals) and relative stability of the cue within an individual over time. One phenotypic character that often shows these characteristics is odor. This is true for many terrestrial invertebrate and vertebrate species; it can stem from variation in a variety of factors, including protein expression, cuticular hydrocarbons, or even phenotypic variation in diet preferences. Knowing who is related and who is not then requires identifying the odor signature of the target individual and either comparing against a self-referent system (this is what I smell like) or against a parental/sibling-referent system (this is what those assumed to be genetically related to me smell like).

Rather than having an olfactory system with genetically controlled hardwired selectivity for specific self or not-self odors, it appears that many species examined in detail to date learn their self odor during early development in a form of olfactory imprinting. Kin recognition through olfactory cues experienced during early development has been demonstrated in fish, amphibians, and mammals.

In mammals, acquisition of odor meaning may begin before birth. The odor of the amniotic fluid is created by a combination of the mother's odor (i.e., signature odor), the mother's diet, and the fetus's odor. Amniotic fluid flows throughout the oral and nasal cavities of the fetus during prenatal life, and olfactory system function and odor-evoked behaviors emerge well before birth in many species, including rodents and humans. Exposure of fetuses to artificial odors (garlic, lemon) in the uterus via intrauterine injections or maternal diet enhances behavioral responsiveness and preference for those odors after birth. Incidentally, amniotic fluid odor is similar to the odor of colostrum, the initial milk produced by mammary glands after birth, and it has been argued that this similarity may facilitate neonate acceptance of the nipple and milk during the early postnatal period (Coureaud et al. 2001). Many mammalian mothers, in fact, lick their ventrum during parturition, which can transfer amniotic fluid from the newly born pups and vaginal area to the nipples, again potentially facilitating the in utero to ex utero transition by the newborn.

Postnatally, the newborn learns that specific odors signal (mean) the mother and siblings through an associative conditioning mechanism (phenotypic matching or familiarization form of kin recognition). The odor to be learned may have both genetic (e.g., MHC gene products) and diet or environmental components, and it may be sufficiently distinct to allow discrimination between the mother and all other postparturient lactating females. In the laboratory, pups can be foster reared, with the birth mother replaced by another lactating female.* In natural conditions, the newborn can generally assume that the female caring for them is the genetically related, biological mother and learning her odor enhances approach and feeding behaviors, as well as subsequent mate preferences. Through association of maternal odor with multimodal sensory stimulation resulting from maternal care (such as warmth, physical contact, milk, and the odor of maternal saliva), the pup learns approach and behavioral activation responses to her odor. According to our definition above, thus, the pup learns the meaning of this novel, complex odor that the mother emits. This type of early olfactory learning has been demonstrated in rats, mice, rabbits, hamsters, and humans, among other mammals, and is sufficiently robust that it can be modeled in the laboratory with classical conditioning. The classical conditioning model of early olfactory learning has proven to be a powerful tool for the study of mechanisms of this behavior fundamental to the survival of the newborn.

More direct evidence that the pup learns the meaning of the maternal odor comes from laboratory work where novel odors were paired with specific unconditioned stimuli meant to mimic different aspects of maternal care. As mentioned above, rat pups learn approach responses to odors paired with a variety of stimuli including warmth, milk, and tactile stimulation. In addition to this generalized approach response, more specific conditioned responses are also learned that are unique to the specific unconditioned stimulus. If an unfamiliar odor is paired with intraoral milk infusions, pups learn to enhance mouthing movements when subsequently presented with the odor. If the odor is paired with tactile stimulation mimicking grooming by

---

* Acceptance of the newborn by the foster mother, like acceptance of the mother by the newborns, is strongly modulated by olfactory cues, as discussed below.

the mother and which evokes behavioral activation in pups, the odor will subsequently evoke conditioned behavioral activation — but not mouthing movements (Sullivan and Hall 1988). Thus, very early in life, odors can acquire specific meaning.

The mother-infant interaction is not solely determined by the neonate, however. Maternal acceptance of the newborn is also influenced by recognition of the newborn's odor. This has been studied most thoroughly in sheep, though excellent work has also been done in various rodents. Human mothers can recognize their infants by smell (statistically speaking) within a short time after birth. Ewes may give birth among a large flock, yet will only allow their own lamb to suckle. Other, nonrelated lambs will be rejected, and this discrimination is based on the mother learning the olfactory signature of its own lamb. This learning occurs rapidly during the initial postpartum period, although there may also be a component of recognizing the maternal self odor signature on the lamb.

As noted above, odors learned during the perinatal period not only influence mammalian mother-infant interactions, but can also influence subsequent mate choice when the neonate attains sexual maturity. One of the fascinating questions derived from this work, and largely unexplored to date, is how, given the enormous developmental changes that occur in the mammalian olfactory system during the postnatal period, are odors experienced during the perinatal period recognized by adults such that mate selection can be influenced by them. What aspect of the memory trace for these odors is sufficiently stable to allow this remarkable feat? In perhaps an extreme example of odor memory maintenance through the lifespan, *Rana* tadpoles exposed to odorants (such as citral) injected into the egg continue to display differential responses to those odors after metamorphosis into adult frogs.

Furthermore, and more directly related to the present discussion, is the question of how the meaning of the odor changes from directing approach for nutrition and maternal care to enhancement of sexual behaviors. Clear demonstrations that odors learned early in life influence mate choice have been made in rats and rabbits. In rats, odors associated with maternal care, or stimuli mimicking maternal care during the first 10 days after birth subsequently enhance male sexual behavior (mountings, intromissions, ejaculations) toward females with that same scent. In rats, the first 10 days appear to be a sensitive period for acquiring odor preferences, because almost any stimulation paired with an odor results in a learned approach response by the pup toward the odor. This includes moderately painful stimuli such as tailpinch and footshock. After postnatal day 10, such stimuli produce learned odor aversions. This apparent confusion by the pup over good and bad (learn to approach odors paired with either milk or tailpinch) appears to be adaptive because, in rats, maternal care in a small crowded nest can result in occasional painful stimulation and the cost of learning an aversion to maternal odor is death. This adaptive confusion has lifelong consequences, however (Regina Sullivan refers to this as "good memories of bad events"), as odors paired with footshock during early life enhance sexual behavior by males toward females in the presence of those odors. The potential relevance to human behavior is currently being explored, but again these results demonstrate how odors can acquire meaning and the robust nature of this association (Freud was meant to be an olfactory neurobiologist).

In mice, phenotypic matching decreases mating with similar smelling (MHC-similar) males. This selective behavior enhances outbreeding and increases MHC variability and individual odor distinctiveness. Once a mate is selected (females show selective choice to live within the territories of MHC-different males), females learn the odor of the male during mating. Association of vaginal distension and the male's odor appears to be sufficient for formation of this odor memory. If the male with whom that female mated loses his territory and a new male approaches within a few days of copulation, the female will abort (fail to implant) the fertilized eggs, thus allowing her to come into estrus sooner and mate with the new male on the block. The neurobiology of this olfactory memory for highly distinct odor signature has been worked out in detail by Keverne and colleagues and is discussed below. However, these data demonstrate that odor meaning can be acquired, in this case mate odor, in a fairly short period of time (courtship and copulation) and have significant physiological consequences.

In addition to phenotypic matching and familiarization for learning kin odors, olfactory recognition of kin can also be based on self-referent matching. In hamsters, young hamsters appear to learn their own odor (armpit effect) and then compare the odor of others against that template to determine kinship (Mateo and Johnston 2000). Cross-fostering studies with hamsters have demonstrated that simply being raised by non-kin does not shift self-identification toward the foster parents, in contrast to what has been observed in phenotypic MHC-matching mice as described above.

A final example provides evidence of cross-modal associations wherein odors seem to acquire the properties of pheromonal stimuli. Sexually naïve male hamsters must have an intact vomeronasal system to successfully mate with a female. Specifically, lesions of the vomeronasal system that leave the main olfactory system intact disrupt mating in sexually naïve male hamsters.

However, the same lesions in sexually experienced males do not disrupt subsequent mating behavior. These results suggest that through mating, female odors detected by the main olfactory system can come to substitute for vomeronasal stimuli. One might speculate this is a similar process as described by Stevenson wherein odors can acquire gustatory qualities in humans, such as the sweet small of cherry odor. In hamsters, odors can become sexy, but only after sexual experience initiated through vomeronasal stimulation.

Food, Foraging, and Homing Odors.   In addition to learning the unique odor signatures of kin and mates, odors can acquire meanings related to food, foraging preferences, and homing behaviors. In contrast to pheromonal cues, which can be under tight evolutionary pressure to align signals with response characteristics of the receiving olfactory system, specific food and homing odors may vary more widely over time and from generation to generation as habitats change or unexpected foraging opportunities arise. This results in a situation in which chemosensory systems may be more effective if they are more widely tuned or adaptable.

For example, as described above, ants have highly selective chemical communication systems where specific chemosensory cues produce specific behavioral responses, such as alarm or aggregation pheromones. They may also have chemosensory biases for certain food or plant odors; however, in some conditions these chemosensory cues related to foraging odors appear to be more plastic, i.e., odor meanings can change with experience. The gardening, leaf cutter ants described above are a good example of this experience-induced shift in meaning.

This example is important because, as noted above, it emphasizes that even in an invertebrate that is often seen as the epitome of physicochemical encoding of odorants (i.e., this physicochemical stimulus evokes this specific behavioral response), adaptive plasticity in odor processing is also possible. Novel odorants can be learned, and as stimulus contingencies change, the associative meanings of those stimuli can also change.

Similar examples are apparent in honeybees that can learn novel odors that signal foraging sites. Similarly to ants, bees have a variety of essentially hard-wired pheromonal circuits and evoked behaviors. However, in addition to these circuits, they are capable of learning to approach specific odors that have been paired with good foraging sites. In fact, these odors can be learned not only by directly experiencing the temporal association between smelling the odor and being at the foraging site, but also by having that scent provided to them during communication with other foragers (Menzel and Muller

1996). Thus, in addition to learning distance and direction to foraging sites during the classical bee communication dance by returning foragers, odor information about the identity of the specific site is also learned. The forager, in a sense, learns that a particular odor means (signals) a good foraging site and transfers that information to other bees in the hive.

In vertebrates, novel odors can gain or lose appetitive meaning in several ways. Simple associative conditioning, in the lab or in the field, is used by many species to learn which odors predict food or danger. Pairing odors with food or with footshock is a standard laboratory method of providing odors with meaning. As in the ants and bees described above, however, social communication can also serve as a method to attach meaning to potential food odors. Work by Geoff Galef and others has demonstrated that social communication between rats can allow naïve animals to learn the potential safety or danger of novel food odors (Galef and Giraldeau 2001). In the poisoned-partner paradigm, one rat is fed a novel food and then injected with lithium chloride to induce illness and gastrointestinal malaise. That rat will subsequently avoid that food. Naïve rats who smell the breath of their sick partner will also subsequently avoid that food, even though the naïve rats themselves never consumed it. Similarly, naïve rats can sample the breath of their conspecifics to learn which novel foods are not associated with illness and thus deemed safe to consume. Although an interesting process for a chemosensory scientist, this process can lead to sleepless nights for pest control professionals attempting to reduce rat populations with poison.

Reversible changes in food and food odor preferences can also occur over short periods because of simple exposure. Sensory-specific satiety in animals is a phenomenon described by Edmund Rolls and others wherein an animal that is fed to satiety on one food, may continue to seek out other foods to satisfy their hunger. Thus, the palatability (pleasantness) of bananas and banana odor may decrease after eating many bananas, whereas the palatability of other foods remains high. In general, responsiveness to and pleasantness of food odors (and food tastes) is modified by hunger and by specific feeding history. Many animals are capable of modifying their diet based on specific nutrition needs, and this in part can be explained by selective changes in sensitivity or reactivity to food odors. Hunger enhances responsiveness to food odors.

Finally, odors are also important for homing and home range recognition for many species. A variety of pheromones are used by animals to mark home territories and nest sites and paths used in homing. In some species of ants for example, pheromonal trails are laid on outgoing foraging trips to facilitate return to the nest. However in many species, olfactory markers of home are learned. One of the more remarkable examples are Pacific Coho salmon described earlier, which return to their natal stream to spawn after spending years in the Pacific Ocean based on learned odor cues. Similar examples of learning the location of the home site have been made in frogs (Hepper and Waldman 1992), pigeons, and mammals. Together, these examples further emphasize that odor meaning can be acquired.

Summary

The examples in this section have been selected to emphasize that olfactory and chemosensory stimuli are critical for a wide range of behavioral functions across the entire microbial and animal world. All of these organisms have some way of detecting and discriminating chemical stimuli, some way of judging relative stimulus intensity, or processing stimuli in such a way that fluctuations in intensity do not disturb stimulus discrimination, and some way of assigning meaning to chemical stimuli.

Several basic tenets of olfactory function can be extracted from these examples. First, in single-celled organisms, these functions are largely hard wired such that the physicochemical properties of the stimulus itself provide discriminative value and meaning. For example, in bacteria, activation of a receptor by the appropriate ligand produces a cascade of intracellular events that result in attraction or repulsion.

Second, most animals, invertebrates and vertebrates, have two modes of olfactory processing, occasionally anatomically separate but frequently intermixed: one mode that depends on the physicochemical nature of the stimulus to encode both identity and meaning and a second mode that is more dynamic and capable of synthesizing novel combinations of odorants into perceptual wholes, including meaning. Each of these processing modes has their own strengths and weaknesses. The more hard-wired mode was presumably shaped over evolutionary time by sexual and sensory selection to bias receivers to specific chemical cues or chemical mixtures used in intraspecific communication, or interspecific predation or predator avoidance. This mode ensures rapid selection of appropriate (though not necessarily reflexive) behaviors in response to specific, relatively stable (transgenerational), signals. The more dynamic processing mode requires additional computation and experience to shape behavioral responses to novel, potentially less predictable stimuli. Today, this odorant mixture signals a good foraging site; tomorrow, a different mixture signals a good foraging site. Learning a novel stimulus, in general, a complex mixture of multiple odorants given the nature of volatiles released by naturally occurring objects, almost invariably involves a synthetic process where the complex mixture comes to evoke a perceptual whole. Conspecific individual A is represented by this complex combination of odorants, whereas individual B is represented by this different, though potentially overlapping, odorant mixture. Learning a new perceptual odor object should generally occur under conditions of attention to the stimulus and/or association of that stimulus with a biologically significant event, i.e., during social interaction, feeding, foraging, homing, predator avoidance, etc. This suggests that the synthetic representation of that perceptual odor object would include components of the context and/or meaning of the stimulus. Without meaning or significant context, the odor object would not have been learned.

The following section takes this interpretation of the behavioral observations described here and attempts to determine their validity relative to what we know about olfactory system neurobiology. Does, or can, the olfactory system function as a dynamic, synthetic processor, incorporating stimulus quality, context, and meaning into a single perceptual odor object?

What Function Implies about Process

The problems olfactory systems must solve to allow odor behavior outlined above—detection of a stimulus against a background, discrimination between similar stimuli, determination of relative concentration, and assessment of meaning—can be solved at the neural circuit level in several different ways, each resulting in specific, predictable advantages and limitations. Perhaps the simplest way to solve all of these problems simultaneously is to have low-threshold, very narrowly tuned receptors and hard-wired central pathways (comparable to the specialist receptors described by Hildebrand and Shepherd [1997]). In this case, if the appropriate stimulus is present, the corresponding receptor and pathway are activated and stimulus-guided behavior can occur. Chemosensory behavior in single-celled organisms such as the bacterium *E. coli* and in *Paramecium* relies on basically this type of mechanism.

A Physicochemically Based System

E. coli require specific nutrients, such as the sugar glucose, and must avoid specific toxins such as phenol or cobalt, and move appropriately within concentration gradients of these substances; for example, either toward higher concentrations of glucose or down concentration gradients away from phenol. *E. coli* detects the presence of glucose via specific glucose-binding receptors located in its cell membrane (Adler 1969). Activation of the glucose receptor activates a second messenger cascade that modifies flagellar movement to increase the probability of long, straight swims. If glucose is not present, or is present in the same or weaker concentrations to which the bacteria have adapted, flagellar movements are modified to increase random movements (tumbling). This receptor-mediated switching between straight swims if relatively high concentrations of glucose are detected, and tumbling if not, results in a random walk up concentration gradients toward the attractant. The attractant is detected and discriminated from other compounds by the nature of the receptor-binding site. No higher-order processing is required. Other compounds within the bacterium's soupy environment are either detected by different receptor proteins with different binding properties, or simply not detected (induce no behavioral consequences) at all.

Thus, in *E. coli*, behavioral sensory discriminations between the attractant glucose and other biologically relevant stimuli depend on the binding affinity of specific membrane-bound protein receptors for the physicochemical features of molecular stimuli. Behavioral discrimination between glucose and many, many other compounds in the *E. coli*'s environment occurs because the *E. coli* have no receptors for those other compounds, and are essentially blind to their existence.

A similar example of olfactory-guided behavior based on the physicochemical properties of the stimulus and mediated by narrowly tuned, low threshold receptors in invertebrates is *Caenorhabditis elegans*. *C. elegans* is a transparent, soil-dwelling nematode approximately 1 mm in length, and uses olfactory cues to regulate movement, feeding, and reproductive behaviors. Specific, identified molecules serve as chemoattractants or repellents, or as intraspecific communication pheromones. The relatively simple nervous system of the *C. elegans* (302 neurons, no more, no less) includes several different chemoreceptor neurons, each of which expresses several different receptor genes (note this is different from most other invertebrate and vertebrate olfactory receptor neurons studied to date where a single receptor neuron is believed to express a single type of olfactory receptor gene). *C. elegans*, like *E. coli*, are fairly limited in their chemoreceptive repertoire. Thus, many compounds that a chemist, dog, or even human could detect in the environment may be missed by *C. elegans*, because the receptors required for recognizing those compounds simply are not expressed. *C. elegans* genetic mutants lacking a specific olfactory receptor gene have a specific behavioral anosmia for the odor ligand normally recognized by that receptor. For example, the *odr-10* gene codes for a receptor protein that recognizes the odorant diacetyl. *odr-10* mutants express a specific behavioral anosmia for diacetyl, with behavioral responses to other odors intact. Thus, *C. elegans* solves the problem of odor detection and discrimination by expressing different receptor proteins that each recognize a specific, biologically relevant odorant molecule. Activation of different receptor proteins produces different odor-guided behaviors.

Yet another example of narrowly tuned chemosensory systems are systems evolved to deal with pheromones. Pheromones are intraspecific chemosensory stimuli that have informational value to the receiver about the sender, such as gender, reproductive status, or kinship. Pheromones are included in the class of chemical signals called semiochemicals and are used by many vertebrates and invertebrates. In many species the precise molecules involved have been identified. In terrestrial animals, pheromones can be either volatile and thus airborne, or large, less volatile molecules. Each of these two types have different characteristics; for example, the large, less volatile compounds provide more long-lasting, more private signals, and the volatile compounds provide a rapid, though short-lasting and less private (easily detected by unintended passers-by) signal, and are thus used in different situations. Aquatic animals also use pheromones, though volatility is less of an issue.

In invertebrates, pheromones may consist of either a single molecule, such as bombykol ((E,Z)-10,12-hexadecadien-1-ol) in silk worm moths, *Bombyx mori*, or more commonly as mixtures of two or more compounds, such as in *Heliothis virescens* moths described above. Recall that, the female *H. virescens* releases a pheromone with a ratio of 1:0.5 of (Z)-11-hexadecenal to (Z)-9-tetradecanal. Varying either the identity of the components or their ratio reduces or eliminates the attractiveness of the pheromone to males of this species. Pheromonal attraction behavior in these moths is mediated by specific olfactory receptors on the antenna that project to the macroglomerular complex in the antennal lobe. In *H. virescens* this glomerular complex is composed of several separate glomeruli, which are highly tuned to the pheromonal mixture components. A small subset of second-order neurons, some of which have dendritic arbors in multiple glomeruli, are highly, selectively tuned to the species-specific sexual chemoattractant mixture, and are much less or totally nonresponsive to other mixture ratios or individual mixture components (Vickers, Christensen, and Hildebrand 1998). This suggests a highly stereotyped processing system, dependent on the physicochemical features of the odor stimulus for evoking behavioral responses.

In further support of a stereotyped encoding of the physicochemical features of moth chemoattractants, *M. sexta* moths, which display pheromonal-guided behaviors similar to the moth discussed above, were subjected to antennal transplants (Schneiderman et al. 1986). In these experiments, male antennae were grafted onto *Manduca* females and allowed to innervate the antennal lobe. Normal *Manduca* females are unresponsive to female chemoattractants, but in response to the odor of tobacco leaves, they display a stereotyped approach and ovipositing behavior. In the transplanted females, chemoattractant pheromones now elicited approach and ovipositing behavior. These results suggest that in this invertebrate system the peripheral olfactory receptors respond to specific physicochemical stimuli, and when excited can activate stimulus-specific behavioral programs. Discrimination of the appropriate stimulus for pheromone-guided behavior from other stimuli is largely driven by the selectivity of the peripheral receptors and the physicochemical features of the stimulus. That is, stimuli with different physicochemical features either bind to different receptors that activate different behavioral responses or are ineffective at binding to any receptors and are thus not perceived. Vertebrates may use a similar physicochemical process for identification of pheromones (Leinders-Zufall et al. 2000). Thus, pheromones or pheromone mixture components can be detected and discriminated from different chemical stimuli by a highly selective peripheral receptor mechanism.

Finally, some species of fish can also serve as examples of olfactory/chemosensory systems that discriminate stimuli largely through highly selective receptors and the physicochemical features of the stimulus. Channel catfish (*Ictalurus punctatus*) and similar species use amino acids as olfactory stimuli to signal the presence of food (Valentincic and Caprio 1994). Exposing a passively resting catfish to a mixture of amino acids can cause searching behavior, as if food is nearby.

However, individual recognition is more complex and requires substantially greater processing capacity. The signal complexity (variation between individuals) required for discriminating one individual from another is substantially greater than that required for discriminating species, sex, or even reproductive status. Similarly recognizing and discriminating novel stimuli, such as food odors and unique home range and foraging site odor cues, present substantial difficulties for the physicochemical labeled-line process that works well for more evolutionarily stable signals. Individual chemical signatures and novel food and homing cues must be learned through experience. This learning not only results in association of meaning with the stimulus, but also teaches that a unique combination of odorants in specific relative concentration ratios represents a single perceptual odor object.

A Memory-based System

As described above, most animals, invertebrates and vertebrates, appear to possess both a physicochemically driven olfactory-processing mode specialized for dealing with evolutionarily controlled, relatively stable odor stimuli, and a more dynamic, synthetic processing mode specialized for novel, less predictable, perhaps more complex stimuli. The latter, memory-based mode begins with odorant analysis by figuratively breaking stimuli hitting the receptor sheet into submolecular features, and then synthesizing them back together again, perhaps along with multimodal and contextual components, to result in a single-odor percept. A memory-based mode such as this enhances flexibility in dealing with novel, variable, and less predictable stimuli, though it requires more processing components and requires experience to maximize its effectiveness.

A memory-based olfactory mode should display several defining characteristics. First, odor discrimination should improve with experience as perceptual odor objects are synthesized. This improvement should be most evident for discrimination of molecularly similar odorants. Second, discrimination of simple odor mixtures from their components should improve with experience, as odor mixtures are synthesized into single-odor percepts unique from their components. Third, and related to the second point, discrimination of odorants from background should improve with experience again as odor objects are synthesized which could help in perceptual grouping needed for discrimination of objects from background. Fourth, one of the consequences of forming perceptual objects from multiple components is that those objects can be more easily recognized even if they are degraded or slightly altered in some way. For example, in vision, a familiar face can be easily recognized even if partially obscured by dark glasses or modified by a change in hairstyle. Less familiar faces may not be recognized under those circumstances. Similarly,* once an odor perceptual object is learned, slight

* Comparisons between olfaction and other sensory systems throughout this section are meant to

changes in components, component ratios, or intensity may be processed in such a way as to allow recognition. This ability to deal with degraded or modified inputs could contribute to perceptual constancy over wide changes in stimulus intensity where it occurs. Finally, experience-dependent formation of odor objects may create multimodal, contextual, or hedonic components of the odor perception and potentially of the information encoded by olfactory system neurons.

Odor Acuity Is Experience Dependency

What is the evidence that olfaction in animals includes a memory-based processing mode? First, odor discrimination of similar odorants (odor acuity) can improve with experience. Christiane Linster, Thom Cleland and colleagues have demonstrated that rats' ability to express successful discrimination of similar odorants depends on the nature of the behavior task (Cleland et al. 2002). For example, in a habituation/cross-habituation task rats had some difficulty displaying discrimination of aliphatic acids differing by a single unbranched hydrocarbon (e.g., acetic acid from proprionic acid) but discriminated longer-chain-length differences well. In contrast, using a generalization task, where acetic acid was paired with a reward and the generalization responses to similar odors determined, rats showed significantly better discriminatory performance. Similarly, Linster and colleagues have demonstrated that the ability to make discriminations between enantiomers can improve with training. Thus, mice will not spontaneously discriminate between enantiomers of limonene or terpinen-4-ol, but within 10 trials of odor discrimination training involving differential association of one isomer with reward and the other nonreward, rats will discriminate (−)-limonene from (+)-limonene and within 20 trials will discriminate (−)-terpinen-4-ol from (+)-terpinen-4-ol. This suggests that experience can change odor acuity and, thus, that odor perception is not entirely based on the physicochemical nature of the odorants.

In a more direct assessment of the role of memory in odor discrimination, Fletcher and Wilson (2002) assessed discrimination of similar ethyl esters in untrained, odor-naïve rats using a habituation/cross-habituation task involv-

---

emphasize similarities in functional outcome, not necessarily similarities in underlying mechanism. Both vision and olfaction have analytical and synthetic properties and, as discussed below, hard-wired processing components and more flexible, dynamic components.

ing bradycardiac orienting responses to odors. In this paradigm, rats did not discriminate between ethyl esters varying by a single carbon but did discriminate between larger differences. The rats' odor acuity was also assessed, using the same habituation/cross-habituation paradigm, but 24 hours after one of the odors had been paired with footshock. These animals, using the same behavioral bradycardiac paradigm as the naïve animals, could now discriminate odors varying by a single-hydrocarbon group. This suggests that familiar, perhaps meaningful odors are more easily discriminated, and that odor acuity is memory dependent. In the final phase of this experiment, additional animals were injected with the acetylcholine antagonist scopolamine prior to the odor shock conditioning to disrupt odor memory and olfactory system plasticity (Saar, Grossman, and Barkai 2001; Linster et al. 2003). Twenty-four hours later (in the absence of scopolamine), these animals showed odor acuity similar to completely naïve animals, again suggesting that olfactory experience and memory are critically involved in odor discrimination of similar odorants.

Olfactory sensory physiology also provides evidence of a strong experience-dependent component to odor processing in both invertebrates and vertebrates. Simple odor exposure can modify spatial patterns of olfactory bulb glomerular activity, temporal patterns of activity in ensembles of olfactory system neurons, molecular receptive fields of olfactory bulb output neurons, synaptic efficacy between the olfactory bulb and piriform cortex neurons, and odor discrimination ability of single piriform cortical neurons. Associating an odor with a biologically significant event such as food or footshock produces a whole host of additional changes in widespread regions of the brain. For this discussion, we will focus specifically on changes in sensory physiology of primary olfactory system neurons as odors become familiar.

As noted elsewhere in this volume, one of the remarkable characteristics of olfactory systems is the odor-specific spatial patterns of evoked glomerular activity across the antennal lobe and olfactory bulb. Each odorant evokes a unique spatial pattern of glomerular activity, presumably reflecting the unique pattern of olfactory receptor neurons activated by the odorant. The most thorough studies of glomerular spatial patterns have been performed using radiolabeled 2-deoxyglucose autoradiography, which involves prolonged exposure to a given odor over about 45 minutes (Johnson, Woo, and Leon 1998). The olfactory bulbs are then removed, sectioned, and exposed to x-ray film for analysis. Although these studies have provided important insight into olfactory bulb functional organization and odor coding, more recent imaging studies suggest that, at least during the first few inhalations of an odor in mice, the spatial patterns are not stable. Rather, on sequential inhalations of an odor, the spatial pattern is refined, becoming more focused as some glomeruli are lost from the activated pattern (Spors and Grinvald 2002). The reader is reminded that behavioral odor discrimination can be achieved in a single inhalation (Uchida and Mainen 2003), thus the shifts in spatial glomerular activity patterns would largely influence subsequent perception or may reflect a component of adaptation. Olfactory classical conditioning can enhance odor evoked activity within the glomerular spatial pattern, without changing the pattern itself, although these changes appear to only occur during early postnatal development (Wilson and Sullivan 1994). Nonetheless, these two series of studies suggest an experience-induced adjustment in spatial extent and/or intensity of glomerular activation in response to previously experienced odors.

Two experience-dependent changes have been observed in odor evoked activity patterns of olfactory second-order neurons, which receive input from the olfactory receptor neurons through synapses within the glomeruli and send their axons to more central circuits. In invertebrates such as locusts and honeybees, it has been argued that temporal synchrony of antennal lobe projection neuron firing is a critical component of odor identity coding by this circuit (Laurent 1996). In the strongest statement of this hypothesis, spatial activity patterns in the antennal lobe are largely epiphenomenal and odor identity is entirely dependent on which projection neurons are coactive and in which temporal order.* The specific synchrony pattern evolves over time during brief, approximately 50-millisecond epochs (observed as beta-frequency oscillations in the local extracellular field potential), thus as the odor becomes more familiar, the representation of that odor changes. The temporal synchrony facilitates odor-selective responding in target cells in the mushroom body. Similar results have been observed in the zebrafish olfactory bulb, where over the course of a 1- to 2-second odor stimulus the temporal patterning of activity within an ensemble of mitral cells changes to enhance the distinctiveness between representations of different odors (Friedrich and Laurent 2001). These results demonstrate that odor coding is dynamic and shaped by experience, even in the invertebrate antennal lobe and nonmammalian olfactory bulb. At present it is unclear if these short-term dynamics in temporal patterns are retained over repeated stimuli, or if each rep-

---

* Thus, for example, orange odor may be encoded by the fact that projection neurons A and B fire synchronously for the first 50 milliseconds of stimulation, then cells A, C, and D for the next 50 seconds, then B, C, and D for the remaining 100 milliseconds of the 200-millisecond stimulus.

etition of the stimulus induces a resetting of the pattern. If the latter is the case then it has been argued that some of the ensemble dynamics in this case may be related to stimulus dynamics in intensity or structure of the odor plume (Heinbockel, Christensen, and Hildebrand 1999; Christensen, Lee, and Hildebrand 2003).

In rats, mitral cell molecular receptive fields can be shifted by odor exposure. Mitral cells have excitatory molecular receptive fields spanning three to five carbons along the carbon-chain-length dimension, generally with a single carbon chain length evoking the largest response (best odorant). In urethane-anesthetized rats, prolonged exposure (50 seconds, approximately 100 inhalations) to an odor with a carbon chain length slightly longer than the best odorant produces a shift in the molecular receptive field toward that exposure odorant (Fletcher and Wilson 2003). The shift is not observed immediately, but rather builds over at least 60 minutes. In some cells, this shift is sufficiently strong to cause the exposure odorant to become the new best odorant for that cell. Given the relatively brief induction period and the relatively short time course between exposure and expression, it seems unlikely that there is a change in olfactory receptor expression or glomerular innervation. Rather, these results have been interpreted as due to a shift in efficacy of intra- or interglomerular synapses, or potentially changes in cortical feedback to the olfactory bulb circuit. The latter hypothesis is appealing because of the extensive feedback to the olfactory bulb from olfactory cortical regions, and because of the well-documented role of centrifugal input on olfactory bulb excitability and odor responsiveness (see Multimodal Odor Perception, below).

Whatever the specific mechanism, an experience-dependent shift in mitral cell molecular receptive fields toward familiar odorant features should result in enhanced representation of those features. Shifts in receptive fields and cortical representations in other sensory systems are correlated with changes in sensory acuity, and thus could be assumed to similarly correlate with, or contribute to, the observed experience-dependent changes in behavioral odor acuity described above. Increased numbers of mitral cells encoding familiar features could enhance contrast of those features from unfamiliar ones, and/or enhance the intensity of or reliability of transmitting information about those features to olfactory cortical neurons.

In addition to changes in projection neuron response patterns, odor exposure and odor learning have also been shown to influence survival of both juxtaglomerular (Woo and Leon 1991) and granule cell interneurons (Rochefort et al. 2002). This experience-dependent effect on interneuron number is odor specific in that cell counts are only affected near or underlying glomeruli specifically activated by the familiar odorant. Increases in numbers or efficacy of olfactory bulb interneurons can have a variety of effects that influence odorant feature encoding. For example, Laurent and colleagues have demonstrated that disruption of inhibitory action of antennal lobe interneurons in honeybees, and enhancement of inhibitory action of olfactory bulb interneurons in transgenic mice impair and enhance odor acuity respectively. Thus, if experience itself can influence interneuron number or efficacy, then odor acuity becomes experience dependent and is memory based.

Finally, several lines of evidence suggest that neurons in the piriform cortex serve as sites of anatomical convergence of multiple odorant features extracted by the periphery and olfactory bulb and that piriform cortical circuitry is ideally suited to learn familiar patterns of input. These patterns may include both olfactory and nonolfactory information, contributing to complex, multimodal odor perceptions, but the discussion here will focus on olfactory input only (see below for multimodal representations).

Linda Buck and colleagues have demonstrated that mitral cells conveying activity from different types of olfactory receptors (and thus, presumably conveying information about different odorant features) terminate in small overlapping patches within the anterior piriform cortex (Zou et al. 2001). Thus, single piriform cortical neurons can receive convergent feature input and therefore begin the process of feature synthesis and formation of perceptual odor objects. However, in addition to convergence of afferent input, the piriform cortex includes an extensive intracortical association fiber system, where a single piriform cortical pyramidal cell may have excitatory associative connections with over 2,000 other cortical cells (Johnson et al. 2000; Yang et al. 2004). It is these associative connections that are hypothesized to be critical for synthetic processing of odor input, and it is these associative connections that are most expressive of use-dependent associative plasticity such as synaptic long-term potentiation (Barkai et al. 1994; Haberly 2001). Afferent connections are also plastic and may contribute to behavioral habituation, adaptation to background, and other forms of odor memory (Best and Wilson 2004).

The intracortical association fiber system has been hypothesized to allow the piriform cortex to function as a combinatorial array, with individual, potentially scattered cortical neurons dynamically synthesizing combinations of odorant features. The Hebbian, use-dependent plasticity of association synapses then allows the circuit to remember previous patterns. The cortical memory adds a significant increase in the power of the olfactory system to recognize and discriminate odors over simple coincidence detection. The cortical memory embodied in modified efficacy of association synapses can allow the cortex to fill in missing components of degraded inputs, conceptually in much the same way that the visual system allows you to recognize a familiar face even if the face is partially obscured from view. Dealing with degraded inputs is critical for a system that must discriminate patterns of overlapping inputs, such as when an odor is presented against a background of odorants. If the target odor object shares submolecular features with the background, then habituation to the background (and thus the overlapping features) could obscure recognition of the target odor. However, once the target odor has been learned by the piriform cortex, habituation to a few features shared with the background could be compensated for by the association fiber synapse memory, and the odor object perception allowed to complete. A similar process could allow enhanced discrimination of familiar odors from each other compared to novel odors. Neural network modeling by James Bower and Michael Hasselmo has elegantly demonstrated the power of the piriform to complete degraded patterns of input. This process, however, can only occur with familiar input patterns, i.e., familiar odors.

An effective visual analogy here is perceptual grouping (fig. 5.3). On first examination, it is unclear which features belong to the target figure and which belong to the background or other figures. After appropriate experience, which could include coherent spatial movement of visual objects or changes in contrast or past experience with the specific object, however, it becomes possible to discriminate, in this case a Dalmatian dog, from the background. The visual system learns which features to group into perceptual objects. Furthermore, once the pattern has been learned, it becomes easy to discriminate the Dalmatian from other objects in the scene and even recognize it when partially obscured. This in brief, is the function of the memory-based olfactory system.

In accord with this model, neurons in the anterior piriform cortex improve at discriminating similar odorants with experience (Wilson 2001b). Using a habituation/cross-habituation paradigm to analyze rat piriform cortex single-unit discrimination of odorants varying in carbon chain length, we have found that prior experience and neural plasticity play a critical role. After a 50-second exposure to a novel odorant such as heptane, piriform cortex single units can discriminate (show minimal cross-habituation to) pentane. In contrast, mitral cells show strong cross-habituation, as would be expected if mitral cells respond to multiple odorants simply because they share a single common feature. Habituate to that feature and responses to all odorants

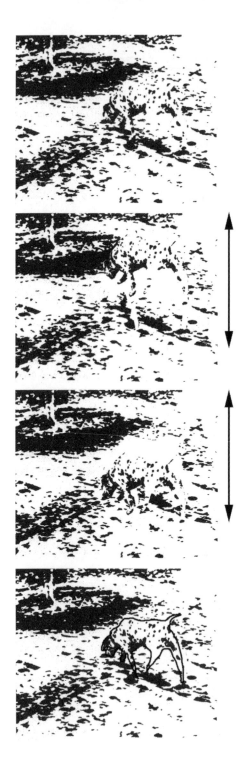

Fig. 5.3. An example of experience-dependent perceptual grouping. Coherent spatial movement of some of the black splotches can lead to grouping of those splotches into a visual object, in this case a Dalmatian dog against a background. Once this grouping is learned, it is easy to see the dog against the background even without the movement cues.

should decrease. But anterior piriform cortex neurons show minimal cross-habituation to similar odorants after 50 seconds of exposure (familiarization). If we assume that the cortical circuit is learning the input pattern for heptane during the 50-second exposure, and it is that learning that allows good discrimination between heptane and other patterns, then disrupting that learning (and its underlying synaptic plasticity) should impair the ability of the cortex to perform perceptual grouping and thus impair odorant discrimination by its neurons. This is precisely what was found. Disruption of normal plasticity by the cholinergic receptor antagonist scopolamine (the same drug that prevents behavioral odor perceptual learning, described above) disrupts odor discrimination by piriform cortical neurons, leaving them functioning largely similarly to mitral cells (Wilson 2001a). These results suggest that without normal synaptic plasticity, the cortex functions largely as a passive coincidence detector and loses its ability to deal with degraded input patterns.*

Odor Mixture Synthesis and Analysis Is Experience Dependent

As described earlier, simple odor mixtures can be perceived either in an analytical manner, as a sum of their individually identifiable parts, or in a configural or synthetic manner where the mixture has unique perceptual properties distinct from the components, which can not be individually identified. Configural processing seems to be the default condition for complex mixtures of more than three or four components. Discrimination of mixtures from their components and, to some extent, analysis of mixtures into components should be improved with experience in animals. This kind of experience-dependent mixture synthesis and analysis would influence both odor olfactory perception and figure-background separation. Given the nature of peripheral odor processing—essentially highly analytical recognition of submolecular features—the ability to synthesize and/or analyze odor mixtures is essentially a perceptual grouping problem. One set of features should be processed as a unitary object different from another, potentially overlapping set of features. Simple recognition of the physicochemical features of odorants hitting the receptor sheet and subsequent anatomical convergence of those features in central circuits is insufficient to account for perceptual

* In this system, scopolamine may actually not reduce synaptic plasticity but rather remove modulatory control of plasticity, allowing synapses to change that should not and thus create interference between patterns. Nonetheless, without normal synaptic plasticity of association fibers, input patterns cannot be learned and synthetic coding and perceptual grouping disrupted.

grouping of mixtures or odors against a background. As memory and experience are added to the system, features that frequently co-occur can be learned and synthesized into unique objects. This can facilitate identification of objects within a mixture or against a background under some circumstances, and in others, when the components themselves are synthesized into a single perceptual object, may reduce mixture analysis.

Prior experience with odor mixtures and their components can affect odor mixture analysis or synthesis in animals. This work has primarily focused on binary mixtures that can be processed either analytically or synthetically, although in some cases more complex mixtures have been examined. As mixtures become more complex, the ability of animals to identify individual components decreases and synthetic processing increases (e.g., invertebrates [Derby et al. 1996], primates [Laska and Hudson 1993]). If spiny lobsters (*Panulirus argus*) are conditioned by pairing a binary mixture of food odorants with a visually aversive unconditioned stimulus, they show very little generalization to the individual components of that mixture, suggesting formation of a unique synthetic percept of the mixture. However, if the lobsters are instead trained in a differential conditioning task where mixture AX is paired with the aversive unconditioned stimulus and mixture AY is selectively unpaired with the unconditioned stimulus (thus components X and Y are differentially conditioned) the animals do generalize between AX and component X and AY and component Y in an analytical manner (Livermore et al. 1997). Thus the context and differential reward conditions can influence whether binary mixtures are treated synthetically or analytically. Similar results have been obtained in the terrestrial slug *Limax marginatus* and the honeybee (*Apis mellifera*).

The *Limax* data are particularly interesting in that they suggest that previous experience with the components alone enhances the ability to identify them in a mixture, whereas mixtures of unfamiliar components are more likely to be processed synthetically (Sekiguchi et al. 1999). These data fit with the model described earlier wherein central circuits familiar with a particular input pattern should be more successful at perceptually grouping and identifying that pattern against a background. In this case, if the central circuits have had experience with odorant A and odorant B individually, it should be easier to identify those components when they are presented in a binary mixture. Conversely, without prior experience with A or B alone, identifying them within a mixture should be difficult, as was the case in *Limax*.

Similar to the behavioral data above, there are good sensory physiology data supporting both configural processing of odor mixtures and the effects of experience on this configural processing. Odor mixtures have opportunities for interaction and thus creation of novel percepts, at the receptor, within olfactory bulb circuitry, and within cortical circuitry. Interaction at any of these levels that results in a spatiotemporal pattern of activity different from that evoked by the algebraic summation of the components presented individually may result in a unique configural perception of the mixture and loss of information about the individual components.

Similar to ligand-receptor interactions in pharmacology, odorants in a mixture can interact at the receptor. Effective ligands for the I7 receptor in rodents include octanal and citronella. Citral appears to act as an antagonist at this receptor, reducing effectiveness of octanal to evoke responses in receptor neurons expressing the I7 receptor. Based on functional imaging of olfactory bulb glomeruli, octanal activates other odorant receptors in addition to the I7 receptor. However, if octanal were presented in a mixture with citral, one piece of the normal octanal spatiotemporal pattern would be missing. This interaction at the receptor, therefore, could disrupt recognition of the octanal as a component of the mixture and result in a unique, synthetic percept of the mixture. Recent work by Leslie Kay and colleagues has, in fact, demonstrated this in rats (Kay, Lowry, and Jacobs 2003).

Components of mixtures can also interact through circuitry in the olfactory bulb or antennal lobe. In both *Manduca* and honeybees, binary odor mixtures can evoke activation of unique glomeruli that are not evident in responses to the components alone. This could be due to interactions at the receptor as above, or could be due to lateral inhibitory and excitatory circuitry within the bulb itself, modulating glomerular activity. In either case, activation of unique, mixture-specific glomeruli results in unique spatiotemporal patterns of projection neuron activity that third-order neurons in the mushroom bodies or piriform cortex must process. This spatiotemporal pattern may or may not include the complete patterns of the individual components and thus, as above, synthetic processing with a unique mixture percept and limited identification of components results.

Finally, the more complex the odorant mixture is, and the more overlapping in features and corresponding patterns of olfactory receptor activation the components are, the more difficult the task of perceptual grouping facing the piriform cortex. We hypothesize that it is the limit in perceptual grouping or pattern recognition abilities of the piriform cortex that places the upper bound on mixture analysis at three components. Beyond that limit, individual component analyses become faulty and odorant mixtures are processed as a single perceptual gestalt.

Although odorant mixture interactions as described above can occur even to novel, unfamiliar odorants, experience can shape cortical mixture processing (Wilson 2003). Single units in the rat anterior piriform cortex can discriminate (show minimal cross-habituation) between binary mixtures and their components, if the mixture is familiar. With novel binary mixtures (less than 20 seconds of experience) anterior piriform cortex neurons are unable to discriminate the mixture from its components. In other words, when mixtures are familiar, cortical neurons treat them as unique objects, different from their components. Without experience, cortical neurons appear to treat mixtures and their components as more similar, consistent with the notion outlined above of neurons simply functioning as coincidence detectors, without synthetic memory or the ability to complete partial inputs. With habituation to a novel mixture, the cortex is unable to complete the degraded patterns that constitute the mixture components, and discrimination fails (i.e., strong cross-habituation). After experience, the cortex has learned to treat the combinations of features constituting the mixture as a unique, complete, synthetic object, distinct from the patterns of features constituting their components.

Odor Perception and Coding Can Be Multimodal and Experience Dependent

One of the remarkable characteristics of the mammalian olfactory system is the extensive centrifugal input providing direct or indirect multimodal information to even the earliest stages of the pathway. The olfactory bulb receives a large feedback projection from the olfactory cortex, as well as modulatory inputs from the cholinergic basal forebrain and monoaminergic brainstem nuclei. In addition to this convergence of olfactory and multisensory activity directly within the primary olfactory pathway, olfactory system afferents project to sites of multisensory, contextual, and hedonic convergence. For example, as noted elsewhere, the olfactory bulb and piriform cortex project to the amygdala, hippocampus, orbitofrontal cortex, and hypothalamus. Thus, within very short pathways olfactory inputs become connected to hedonic, contextual, gustatory, and feeding centers. Single neurons in the orbitofrontal cortex of primates express both olfactory and gustatory response properties. In rats, single neurons in the amygdala respond to both olfactory and aversive somatosensory stimuli. Furthermore, in rats, neurons in the piriform cortex respond to olfactory stimuli and nonolfactory components of complex tasks, including stimuli predictive of odor sampling and reward consummation. This early multisensory convergence combined with the role of memory in odor processing, may enhance the inclusion of nonolfactory components into odor object percepts, thus creating odor percepts with inextricably linked hedonic, contextual, and multisensory features.

An interesting example of this is in the work of Jeanne Pager and colleagues. They reported that rat olfactory bulb unit responses to food odors were modulated by the hunger or satiety of the animal (Pager 1978). Thus, olfactory bulb units were hyperresponsive to food odors if the rat was hungry, compared with when the animal was satiated. The magnitude of responses of the same cells to novel nonfood odors were not similarly dependent on the internal state of the animal. This suggests that meaning of the odor influenced how it was processed, even at the very earliest stages of the olfactory system. If the animals had a novel nonfood odor associated with food when they were young, then responsiveness to that odor became under the control of hunger when the animals were adult. This suggests that past experience shapes the meaning (hedonic, contextual components) of an odor and that those components become tied to the physicochemical feature representation.

Similarly, both olfactory bulb mitral cell and piriform cortical cell single-unit responses to odors can be shaped by previous associative experience. For example, the piriform cortical single-unit response to an odor associated with reward is different than the response to novel odors (Zinyuk, Datiche, and Cattarelli 2001), suggesting that these cells may not only encode the physicochemical nature of the stimulus but also its nonolfactory associations.

As noted above, centrifugal inputs to the olfactory system are the primary source of information concerning behavioral state and biological significance. Cells in both the olfactory bulb and piriform cortex in rats respond in a variety of non-odor-sampling components of behavioral tasks. For example, Schoenbaum and Eichenbaum (1995) have demonstrated that neurons in the piriform cortex respond not only to odor sampling, but also during approach to the odor-sampling port, approach to the water reward-sampling port, and during reward consumption. Some of this apparently non-odor evoked activity could reflect changes in sensitivity of the olfactory system with changes in arousal, and thus are not true multisensory responses. Thus, for example, activation of the noradrenergic nucleus locus coeruleus, as might occur during arousal or vigilance, can increase mitral cell responsivity to olfactory nerve input (Jiang et al. 1996). In this case, therefore what might appear to be a multisensory response to task contextual stimuli could actually be a response to background odor cues that is unmasked by changes in nor-epinephrine release in to the olfactory system. Activation of noradrenergic β-receptors has also been shown to reduce activity-dependent depression at the mitral cell–piriform cortical pyramidal cell synapse that has been linked to cortical odor adaptation (Best and Wilson 2004). Schoenbaum and Eichen-baum's data however, suggest that different cortical cells are responsive to different multisensory aspects of the task which argues that not all of this activity is simply arousal mediated, and allows for the formation of complex, multisensory and contextual representations of odor objects (Haberly 2001). Network modeling and neuroanatomical analyses of association fiber connections suggest that the posterior piriform cortex may be most involved in this multisensory aspect of odor percept formation.

The other important component of multisensory processing of odors is the nature of olfactory structures beyond the piriform cortex, including the amygdala, orbitofrontal cortex, and hippocampus. Extensive reviews exist of these three structures and their role in memory, emotion, and perception, thus only a few examples specific to olfaction will be included here. Single units in the lateral nucleus of the amygdala respond to both odor stimulation and footshock in rats (Rosenkranz and Grace 2002). Single neurons in the amygdala can thus serve as convergence sites for odors and their biological significance and presumably contribute to the hedonic aspects of odor percepts. Neurons in the cortical nucleus of the amygdala project back to the olfactory bulb, thus providing another opportunity for state-dependent modulation of odor coding as early as the second-order neurons.

The orbitofrontal cortex receives direct and indirect projections from the piriform cortex, as well as nonolfactory inputs including gustatory and somatosensory inputs. The orbitofrontal cortex has been shown in both rats and primates to include multisensory neurons, for example, single neurons that respond to both odors and tastes. Orbitofrontal cortex neurons, similar to piriform cortex neurons, also respond to many nonodor aspects of behavioral tasks, such as during approach to odor or reward-sampling ports or reward consumption. Furthermore, the response properties of these neurons can be shaped by past experience (Rolls, Critchley, and Treves 1996). As in the piriform cortex, therefore, and perhaps to an even greater degree, opportunities exist for single neurons to acquire complex, multisensory representations of odor stimuli.

Finally, the hippocampus has long been recognized as a site involved in memory and representations of complex stimulus relationships, including representation of spatial maps and representations of simultaneously present stimuli. Gary Lynch and colleagues had hypothesized in fact that the combinatorial array circuitry begun in the piriform cortex is extended in the hippocampal formation to assist in synthesis of complex odor mixtures into perceptual odor objects (Lynch 1986). Damage to the hippocampal formation and/or its afferents impairs performance in odor memory tasks requiring comparisons of odor cues presented simultaneously. The hippocampus receives extensive multisensory inputs and thus could also serve as an important site of formation of higher-order multisensory odor memories.

Summary

This section argues that olfaction can be functional in either a physicochemically driven mode, largely hard wired to detect specific physicochemical features of stimuli and drive labeled line central pathways, or a memory-based mode, relying on a highly analytical periphery and experience-dependent, synthetic central processing of perceptual odor objects built of those peripherally extracted features.

The parallels between the two olfactory processing modes described here (physicochemical and memory based) and processes used in visual systems is striking. In invertebrate predators such as dragonflies and in some vertebrates such as frogs and toads, visual systems include feature-detection systems similar to the physicochemical olfactory mode. Toads, for example as described above, have worm detectors as a component of their relatively simple visual systems wherein visual stimuli with the appropriate spatial characteristics (long and thin and moving parallel to the long axis) evoke a worm-catching-and-consuming reflexive behavior. Thus, this specific spatial stimulus that matches an internal template activates a highly adapted response, which, away from scientists with projectors and fake visual stimuli, helps feed the toad. This is an unlearned response to a specific pattern of receptor and central activation, very similar to the physicochemically driven olfactory processing mode. Other animals have similar visual detection systems, such as fly detectors in dragonflies.

More complex visual behavior, however, requires more complex and flexible processing. A human visual system would probably not be fooled into thinking (under normal circumstances) that a pencil is a worm. Our visual system begins by breaking down a visual field into tiny bits and then puts the bits back together again into perceptual objects. Innate or learned biases and context may influence the synthetic process (its easier to see, or imagine we see some kinds of objects, such as faces, than others), but the breakdown into components and reassembly into objects enhances the flexibility of our visual perception. However, the synthesis of visual objects in the higher-order visual cortex requires experience.

This process is quite similar to that described for the memory-based olfactory processing mode, although there are important differences. Primary among these differences is that once an odor perceptual object is learned from a complex mixture of components, perceptual analysis of the mixture into its components is not possible, while in vision, even after a face has become very familiar we can still describe its individual features. This difference between analytical abilities in vision and in memory-based olfaction may be a consequence of the lack of spatial dimension in the odor stimulus that is a primary feature of the visual stimulus. For example, being able to sequentially scan a visual image may greatly facilitate the analytical process in a way that is unavailable to the olfactory system.

Alternatively, the inability to analyze learned odor objects may simply reflect a limitation of the processing capability of the system. A similar limitation appears in the visual system of some desert ants. These ants learn to home toward the nest based on visual cues acquired on earlier trips. Thus, for example, an ant may learn that the nest is surrounded by a triangle of three stones of a particular size. The ant appears to learn this visual landmark as a template in its visual system, not as a complex of different visual features. If the ant learns the landmark with the upper half of its compound eye while the lower half is experimentally occluded, it cannot recognize the pattern with the lower half of its eye after the occluding mask is moved (Wehner, Michel, and Antonsen 1996). Furthermore, it successfully finds its nest if the stones are moved more distant from the nest, as long as the size of the stones is increased proportionately to the distance to create the same size visual pattern on the eye. These results suggest that the ant learns a visual perceptual object (in this case the landmark around its nest) as a template, and once learned, it cannot be analyzed into its components. This is, at least superficially, similar to the case in memory-based olfactory processing.

Constraints Imposed by This Solution

The two-mode olfactory system described here maximizes fidelity of responses to highly evolved chemical cues such as odors and maximizes flexibility of response to novel odors composed of complex mixtures. As described elsewhere, this combination of a hard-wired processing mode driven by specific stimulus features, and a memory-based synthetic processing mode capable of responding to novel stimuli, is a common feature of many sensory systems, including visual and auditory systems in many species. At two extremes, there may be examples of organisms with only one or the other system. Bacteria, for example, have the physicochemical processing mode alone. (In more complex organisms, it is important not to confuse the dichotomy of a physicochemical processing mode and memory-based processing mode as necessarily implying a particular anatomical separation.)

Note that a single odorant may be processed by both modes. Thus, pheromone molecules can be recognized by the physicochemical mode as complete stimuli, evoking activation of a specific central pathway. However, features of those same molecules may be processed within the memory-based mode.

Together, these two modes produce a powerful tool for analyzing chemical stimuli. However, they also imply or impose some constraints on processing. These constraints are beyond those already outlined of the individual systems themselves, such as the lack of flexibility and potentially limited response breadth of the physicochemical system. Here, we will focus on potential or evident constraints on the memory-based mode. These constraints include or may include limitations in odor mixture analysis, confusion in multimodal odor perceptions, and biases in synthesis of odor objects.

Limitations on Identifying Parts

First, although the memory-based processing mode begins with feature analysis, it rapidly starts the process of synthesis, and it is that memory-based synthesis that results in synthetic perceptual wholes. This synthetic processing limits conscious or behavioral analysis of mixtures. Behavioral identification of components in complex mixtures and, in some cases, even binary mixtures is highly limited. We hypothesize that this is the case because odor discrimination depends on piriform cortex output, and the piriform cortex functions as an experience-dependent combinatorial array whose output obscures the identity of individual component inputs. Familiar or highly distinct components will be easier to extract from simple mixtures because of the ability of piriform cortical circuitry to recognize even partially degraded or obscured familiar input patterns. However, if the component odorants are highly similar and thus produce overlapping inputs, or if there are a large number of component odorants, the cortical circuitry will be unable to identify the components, but it will still produce a unique perceptual whole.

Most studies in animals, namely bees, lobsters, and rats, have been confined to studying generalization between single odorants and mixtures. Relatively few studies have examined generalization to multicomponent mixtures. For generalization to binary mixtures the findings from different phyla are similar, namely that the experimental task primarily dictates whether the animal treats the odor mixture elementally or as a unique configuration. For example, in vertebrates, rats tested on a habituation paradigm treat mixtures of similar smelling odorants as unique entities, in that they show responding when subsequently exposed to the mixture elements, but *not* when dissimilar odors are mixed together (Wiltrout, Dogra, and Linster 2003; experiment 1). Thus under these conditions animals treat similar smelling odors mixed together as unique entities and dissimilar smelling odors mixed together as two discrete components. However, under different conditions, namely those pertaining to a blocking experiment, the similarity or dissimilarity of the components used in the blocking mixture has no effect on the demonstration of blocking. That is, here, animals must perceive the elements in both the similar and dissimilar mixture for the blocking effect to occur (Wiltrout, Dogra, and Linster 2003; experiment 2).

This interpretation of blocking experiments is not universally shared. Pearce (1997) has argued that configural theories may also predict blocking and that summation may provide a better test to distinguish between elemental and configural accounts. To our knowledge this has not yet been tested in rats using odor mixtures in a sensory preconditioning procedure. However, in honeybees it has and they *can show* summation indicative of elemental processing (Hellstern, Malaka, and Hammer 1998). On the other hand bees can also show evidence of configural processing, in that optimal acquisition of an odor mixture in a sensory preconditioning paradigm occurs after just one exposure. Elemental theories predict that multiple exposures will enhance this further, but they do not (Muller et al. 2000). Thus under certain conditions the components of a binary mixture may be identifiable, but not under other conditions.

Similar results have also been obtained in lobsters. Livermore and colleagues (1997) found that when a lobster was aversively conditioned to the odor mixture AX, no generalization to A, X, AY, or Y could be obtained. However, when animals were aversively conditioned to the odor mixture AX and the odor compound AY was used to signal the absence of aversive reinforcement, generalization from AX to X and from AY to Y was observed. Again, different experimental conditions led the animals to treat the odor mixture as either a unique configuration or a set of elements.

Whether animals, like humans, have limitations on the number of components in a mixture that they can identify, has not been extensively explored and only three studies address this question. Derby and colleagues (1996) examined generalization in lobsters from: (1) an aversively conditioned odor A, to single odorants (A) and mixtures (AB, ABC, ABCD, ABCDE, ABCDEF, ABCDEFG); (2) an aversively conditioned odor mixture ABCD, to single odorants (A) and mixtures (AB, ABC, ABCD, ABCDE, ABCDEF, ABCDEFG); and (3) an aversively conditioned odor mixture ABCDEFG to the same stimulus set described in conditions 1 and 2. They found that there was little generalization from A to the other mixtures. When the ABCD mixture was the conditioned stimulus, generalization to A occurred, but not to any other mixture. When the seven-component mixture was the conditioned stimulus, generalization was obtained to ABCD, ABCDE, ABCDEF, but not to any other single compound or mixture. Obviously the results from this experiment are not directly analogous to the human studies reported in the next chapter, but these findings do show that generalization was typically limited, suggesting that under these conditions (an important caveat as indicated by the experiments described earlier) the odor elements were not readily detected. On this basis then, lobsters do not appear to be that different from humans.

Similar results have also been obtained in squirrel monkeys (Laska and Hudson 1993). Three animals were trained to discriminate a twelve-component odor mixture from carvone. After learning this discrimination, they were then tested to see whether or not their ability to respond to the S+ (the twelve-component odor mixture) would be retained, when various sub-mixtures were substituted for the carvone S−. When three-component mixtures were used (with the components drawn from the twelve odors used in the twelve-component mixture) the monkeys were readily able to select the food reinforced S+. The same held true for six-component mixtures. Performance largely dropped to chance level, however, when 9-component and 11-component mixtures were used. These results suggest that squirrel monkeys treat the 3- and 6-component mixtures as discrete objects because little generalization occurs between them and the 12-component mixtures. Only when the mixtures differ by between one and three components do the animals generalize and thus fail to learn the discrimination.

The most compelling data are from a study using rats (Staubli, Fraser, et al. 1987). These animals were required to learn that the odor mixture ABC indicated the presence of food in one arm of a radial maze. The same animals were also presented with another odor mixture ABD, which indicated the absence of food in another arm of the maze. Once these relationships were acquired animals were then trained with a new set of odors. In this retraining phase odor D now predicted food in one arm of the maze and odor C its absence. The extent of generalization from a mixture to its component is evident here by the degree to which it interferes with learning of the new relationships. Under the conditions described above, the animals acquired the new relationships as rapidly as the old ones, indicating little interference and thus little generalization between the mixtures and their elements. Thus under these conditions the animals did not identify the mixture components. Several other conditions were also tested. When the initial training odors were binary mixtures, some generalization was obtained, suggesting that in this case the animals could detect the components. Not surprisingly, when a condition using four-component mixtures was used, no generalization was found. Yet again, the earlier studies with binary mixtures demonstrate that the experimental conditions can affect whether the animal treats the mixture in a configural or elemental manner. Nonetheless, these findings in rats and those in lobsters suggest that they too are poor at identifying the components in more complex mixtures.

Constraints on Forgetting

Many animals show impressive long-term retention of olfactory material. Rats, trained to discriminate 36 odor pairs, were able to successfully discriminate odors drawn from this set 24 hours after training was completed (Slotnick, Kufera, and Silberberg 1991). Longer retention intervals for smaller sets have also been observed in rats, with 5 of 7 animals retaining an olfactory discrimination between a pair of odors when tested 53 days after training. Mice show similar abilities. When trained with larger stimulus sets (16 odor pairs), they were able to retain successful discrimination between pairs drawn from this set up to 4 weeks after training (Larson and Sieprawska 2002). On simpler problems, where one odor is paired with sucrose and another with water, mice spend more time digging in a cup scented with the sucrose-paired odor, even when this test is given 60 days after training. In fact performance in these animals was considerably better than mice tested 24 hours after testing was complete (Schellinck, Forestell, and LoLordo 2001). Similar findings have also been obtained using different discrimination procedures, with simple two-odor problems being successfully retained by mice for at least 32 days (Bodyak and Slotnick 1999). Other mammals also show the same general pattern. Rabbit pups, briefly exposed to juniper scent prenatally, demonstrate a preference for juniper when tested one month after exposure (Bilko, Altbacker, and Hudson 1994). Guinea pigs retain a memory of another individual's urine odor, following a brief exposure, when tested 4 weeks later (Beauchamp and Wellington 1984). Likewise, hamsters habituated to the flank scent of a male conspecific show more interest in the smell of a novel male conspecific when tested 21 days later. Finally, Laska and colleagues (2003) trained three spider monkeys on an odor discrimination problem and then tested the retention of this discrimination 4 weeks later. The monkeys still successfully performed following this interval. In sum, mammals appear well able to retain olfactory information for periods extending to several weeks, even when the stimuli are drawn from initially large sets.

Few animal studies have directly examined extinction or interference procedures. Laska, Alicke, and Hudson (1996) trained several squirrel monkeys on an odor discrimination problem. During the interval between the original training period and the retention test (conducted seven months later), the animals were trained with a new odor discrimination problem during the last 15 weeks of the retention interval. Although not expressly designed to investigate interference, the rate at which the monkeys learned the new odor discrimination problem provides some insight into the degree to which the old learning interferes with the new learning. The monkeys both acquired the new problem as rapidly as they had the old one, indicating little proactive interference. Moreover, their performance on the retention test of the old learning conducted seven months after initial training revealed that the old problem was at least partially retained, as it was rapidly relearned. These findings suggest little proactive or retroactive interference for this type of odor learning.

Sensory preconditioning is the animal analogue of odor-odor and odor-taste learning in humans (see chapter 6). Unfortunately most sensory preconditioning experiments have not explored the effect of extinction on odor-odor associations, but some data are available for odor-taste associations. These data reveal a dissociation between what appear to be two different forms of odor-taste learning. Harris, Shand, Carroll, and Westbrook (2004) found that rats exposed to an almond-sucrose mixture rapidly acquired a preference for almond-scented water after training. However, the resistance of this preference to extinction is mediated by the motivational state of the animal at test and during acquisition. Rats trained and tested while sated do not show extinction, that is, a preference for almond-scented water over water alone is retained following exposure to almond-scented water unflavored by sucrose, posttraining. Rats trained food deprived or tested when hungry do show extinction, suggesting that animals can acquire either (or both) an odor-calorie association which is *not* resistant to extinction and an odor-sweet taste association which is. The latter finding is remarkably similar to that observed in humans where the dependent variable is odor sweetness ratings. This too is presumably an odor-taste (rather than calories) association, and this too is resistant to extinction.

Honeybees also show a remarkably similar effect (Abramson et al. 1997). Following proboscis conditioning using a sucrose-paired odor, extinction is readily obtained when only the odor is presented, but when the odor and sucrose are presented in an unpaired format, after training, the animals continue to respond to the odor. This is similar to the results obtained in rats because, when the sucrose is presented, the animals are allowed to consume the sucrose for three seconds. Thus, during the unpaired extinction procedure animals can be considered as sated, unlike the extinction condition in which the odor is presented without any sucrose feeding.

The degree to which odor learning is context specific has also been explored, but again in a rather limited way. Humans may be relatively context insensitive with odor-learning phenomena, although little direct evidence is yet available. Rats trained on an odor habituation task to a male conspecific's odor show as much interest to a novel conspecific's odor when tested in a new context, as they do when tested in the original training context, suggesting that retention and recollection of this material is relatively independent of where it was acquired (Burman and Mendl 2002). Similarly, sensory preconditioning using almond odor in saline is context insensitive, because testing preference for almond under conditions of salt deprivation can manifest both in the training context and in a novel context as well (Westbrook et al. 1995).

Although the literature on forgetting is limited in animals, what there is appears similar to findings in humans. Animals can retain olfactory information for long periods, this information can be retrieved in contexts different from the one in which it was learned, and the information, at least some forms, is insensitive to extinction or interference. Information that is sensitive to extinction-like procedures appears to be related to calorie learning, in other words this allows animals to recalibrate the usefulness of an odor as a predictive cue. Recalibration would appear to have little value in information that is insensitive to extinction and to change in context.

Redintegration

Another constraint, and largely a function of the first, is that odor perceptions will include integral multimodal components, and, in the correct circumstances, a reduced subset of these various components may be sufficient to recall the whole (redintegration). Given the synthetic and experience-dependent nature of odor processing described, and the opportunities for multimodal, contextual, and hedonic input to that processing, odor perceptual objects will contain nonextractable multimodal, contextual, and hedonic components. The clearest example of this comes from human introspection and research demonstrating sweet odors. However, animal behavior also suggests strong ties between odors and the associations and learned outcomes. Odor memories may be so robust because the associated memories are components of the odor perception itself.

However, there is currently little evidence available in animals to suggest that a part can recover the whole. One possible line of evidence emerged from bulbar lesion studies, in which an odor's 2-deoxyglucose (2-DG) focus was selectively lesioned. Such lesions have little effect on subsequent ability to discriminate that odor—propionic acid (Lu and Slotnick 1994)—suggesting that the remnant portion of the pattern generated by the glomerular layer is sufficient to activate a representation of the original odor and thus an appropriate behavioral response. This interpretation appears unlikely, however, because animals who have never been exposed to propionic acid, yet who have medial bulb lesions that excise the primary 2-DG focus, are as quick in learning a propionic acid discrimination task as animals that have had no lesion or sham surgery (Slotnick et al. 1987, 1997). These latter findings are encouraging for a different reason, because they suggest that it is the bulb's overall output, rather than output from specific receptor types located on one area of the bulb, that is important.

Finally, we expect that biases may exist within the memory-based mode such that some feature combinations or receptor inputs may be easier to learn or more influential in forming associations than others. Just as there are evolutionarily controlled, biological biases in tuning of the physicochemical mode, there may be evolutionarily controlled, biological biases in feature synthesis within the memory-based system. This may be especially true when including multimodal or hedonic components to the odor perceptual object. For example, olfaction and taste appear to have a special relationship wherein the combinations provide a unique perception called flavor. This relationship may be similar to the special relationship between vision and audition in many animals where spatial maps of visual object location and of auditory object location physically overlap in the superior colliculus, allowing sounds to help orient our visual system. The interaction between these systems can be more than just simple convergence, but actually instructive wherein the visual input to the map can be used to calibrate the auditory map (Feldman and Knudsen 1997). Both the olfactory-gustatory and visual-auditory relationships represent biological biases in multimodal interactions.

Conclusions

For many, perhaps most animals, olfaction appears to be processed in two different modes. One mode functions as a hard-wired, physicochemical detector, driven by specific chemical stimuli capable of evoking stereotyped, though not necessarily reflexive behaviors. This processing is under evolutionary selection pressure to deal primarily with communication signals such as pheromones. The stimuli may consist of single molecules or highly specific, invariant mixtures. This physicochemically based processing mode allows rapid, low-threshold responses to relatively predictable stimuli, with precise, consistent meanings, that have been stable over evolutionary timescales. The primary disadvantage of this mode is its necessarily limited breadth and flexibility in responding to novel stimuli. In contrast, the second processing mode functions in a memory-based, synthetic manner involving peripheral, submolecular feature extraction and central experience-based synthesis of those features into perceptual odor objects. This mode allows recognition of less predictable, potentially more variable and dynamic stimuli whose behavioral meaning may change with experience. The disadvantages of this mode include required additional processing circuitry and limitations of perceptual analysis of odor objects once they are learned.

Individual stimuli may be processed in both modes. These two modes need not be anatomically segregated. Although in mammals the physicochemically based system fits more closely with the vomeronasal-accessory olfactory system and the memory-based system fits more closely with the main olfactory system, the main olfactory system may include physicochemically based subunits. The mammary pheromone in neonatal rabbits, 2-methylbut-2-enal, is an excellent example of a physicochemically based process producing stereotyped behavior, yet being processed by the main olfactory system (Schaal et al. 2003). In addition, the two processing modes may directly interact with, for example, the physicochemical mode functioning to instruct the memory-based mode. Chemosensory control of male hamster mating provides an example of this instructive role. Similarly, association of novel odorants with pheromones in neonates can provide meaning (and thus modify subsequent behavior) to those novel odorants (Sullivan, Hofer, and Brake 1986).

The memory-based processing mode provides a powerful method of dealing with unpredictable, dynamic, highly complex mixtures experienced against similarly complex chemical backgrounds—precisely the conditions under which much of olfactory behavior occurs. In addition to allowing for experience-based synthesis of complex mixtures into unique perceptual objects, the cortical circuitry on which the memory-based processing relies also enhances stability of responses to degraded inputs (Barkai et al. 1994). The ability to recognize partially degraded input patterns enhances recognition of familiar objects from highly similar patterns, enhances detection of those objects against complex backgrounds, and may contribute to perceptual constancy with variations in intensity. The experience-dependent synthesis of odor objects appears to also create complex, multidimensional odor perceptions, which may include inextractable contextual and multimodal components. Synthetic odor objects of complex odorant mixtures become highly distinct percepts with integral learned meanings.

As opposed to the physicochemical mode in which a single component within a complex mixture can be detected and drive behavior, synthetic memory-based processing obscures perception of mixture components. This can have the effect of broadening the information content of an olfactory scene. In physicochemical processing, the presence of odorant A evokes a particular behavior, potentially regardless of the other components of the olfactory scene. In memory-based processing, odorant A may have one meaning if sensed alone or contribute to any number of distinct percepts depending on the combination of other odorants with which it is mixed. This combinatorial processing, while limiting mixture analysis, dramatically enhances the information available to be extracted from the chemical environment.

The combination of physicochemical and memory-based olfactory modes of processing occurs to some extent in most invertebrate and vertebrate species examined to date and has close conceptual parallels in other sensory systems. The presence of dual processing modes in olfaction across the animal kingdom and across sensory systems suggests this is a powerfully adaptive way of dealing with information processing in general.

6

Odor Quality Discrimination
in Humans

The Function of Human Odor Perception

One way to assess the function of olfaction in humans is to examine the consequences of its loss. Humans rely primarily on vision and audition, and so
the sense of smell assumes far less importance relative to many other animals.
Thus, although blindness or deafness are serious disabilities, anosmia exerts
more subtle effects. These include (1) vocational problems where the sense
of smell is required professionally (e.g., chefs, bakers, firemen, etc.; Callahan
and Hinkebein 1999); (2) depression, especially with parosmia (Deems et al.
1991); (3) weight loss through decreased appetite (Deems et al. 1991); (4) safety
issues concerning the identification of smoke, natural gas, and other such
odorants (Mann 2002); (5) problems with personal and food hygiene (Callahan and Hinkebein 1999); and (6) reduced libido (Costanzo and Zasler 1991).
Notwithstanding these practical consequences, the aesthetic loss of no longer
being able to enjoy eating and drinking, where olfaction is the primary sense,
and the inability to experience pleasant natural odors is reflected by lower
quality-of-life scores reported in anosmic individuals (Miwa et al. 2001).

The effects of anosmia point to a number of functional roles for human
olfaction, namely, ingestion, disease avoidance, sex, and hazard warning.

Each of these functions is examined in turn with the aim to understand what the olfactory system needs to accomplish to fulfill that function. Finally, the issue of perceptual expertise in olfaction is touched upon as this too falls within the bailiwick of function, albeit of a highly specialized sort.

Ingestion

Mammalian olfaction is uniquely configured, in that there are two discrete means of stimulating the same set of receptors. The first, sniffing or orthonasal olfaction, is clearly experienced as a distinct sensory channel relative to vision, touch, and hearing. This is suggested both by common experience and by the general tendency to identify the sense with the sensory channel (Abdi 2002). The second route is retronasal olfaction. This involves the diffusion of volatiles during eating and drinking, via the nasopharynx, to the olfactory receptors. Here too, the sense and sensory channel are perceived as one, so that many participants mistakenly regard the sensations evoked by food or drink in the mouth as taste, not smell (Murphy and Cain 1980). Several lines of evidence suggest this. First, patients reporting loss of their sense of smell typically report loss of smell *and* taste, even though their sense of taste is usually intact (Deems et al. 1991). Second, healthy participants are often surprised at the dramatic effect of pinching their nose while eating or drinking (Stevenson 2001a). This simple expedient prevents the diffusion of volatiles via the nasopharynx, leaving just the taste and somatosensory components of flavor (Lawless 1996). Third, the tendency to treat taste and smell as one sense during ingestion is reflected linguistically. Rozin (1982) found that, of the ten languages he surveyed (English, Czech, French, German, Hebrew, Hindi, Hungarian, Mandarin, Tamil, and Spanish), none had terms distinguishing between the olfactory and taste components encountered during ingestion.

A variety of functions are subsumed to the olfactory system during ingestion. Prior to ingestion, these functions include identifying stimuli that are foods, identifying whether they are fit to eat and whether they are likely to be pleasant. Clearly, such functions do not occur without the participation of the other senses, especially vision. Nonetheless, as the anosmia literature indicates, detecting rotten food and judging palatability, in particular, may be adversely affected if olfaction is absent. Once food (or drink) is placed in the mouth, the olfactory component of flavor again contributes to the decision as to whether or not to ingest. In addition, olfaction plays an important role in the regulation of intake, mediated primarily by motivational changes that manifest as alterations to the odor's hedonic attributes. Each of these functions is now explored in more detail.

Children learn relatively early the distinction between foods and non-foods, a process that does not initially appear to be under the control of the sensory properties of the stimulus (Rozin et al. 1986). This changes with age, such that the stimulus properties, especially smell, come to govern whether an item is classed as food and thus fit for ingestion (Fallon and Rozin 1983). In adults, there is a marked reluctance to try novel foods, expressed as a general but mild expected dislike for all new flavors (Pliner and Pelchat 1991). Elisabeth Rozin has suggested that this tendency to only consume familiar flavors may dictate culinary style, given that many food cultures can be reduced to a set of flavor principles (E. Rozin, S. Rozin, and E. Rozin 1992). For example, tomato and oregano in Italian food, or garlic, cumin, and mint in Northeast African food, typify the flavor principles at work in these cuisines. Flavor principles then may allow the introduction of new foods, by flavoring them with a combination that is already familiar, thus rendering the food as edible and bypassing the neophobic response (Rozin 1978).

The expected palatability of food again involves more than one sensory channel. For some foods, smell is important in determining whether the food is ready to eat (e.g., ripeness). Not only is this controlled by knowledge and experience with that particular food type (i.e., hung venison), but also by immediate and delayed consequences from eating the food on previous occasions. Experiences that act to enhance palatability in this way include the food's caloric density (e.g., Booth, Mather, and Fuller 1982) and protein content (e.g., Gibson, Wainwright, and Booth 1995). Those acting to reduce it include whether the food previously made you sick (e.g., Cannon et al. 1983). Once the food is prepared and eating starts, olfaction again serves to detect whether the food should be eaten—the last chance before it is incorporated.

The final function of olfaction in this context is in its contribution to governing meal size. This occurs through two principle mechanisms. The first, sensory-specific satiety, is particular to the type of food being eaten and results in a selective reduction in palatability for that food (Raynor and Epstein 2001). This occurs independently of caloric load, so that intake can be controlled prior to the delayed signals resulting from ingestion. The second, a more delayed effect based on caloric load, renders food odors, in general, as less pleasant postingestion, reaching a peak shortly after a meal and lasting for a number of hours beyond (Duclaux, Feisthauer, and Cabanac 1973). These mechanisms, although not the only ones governing intake, act via an odor's hedonic attributes.

The subjective experience of olfaction during ingestion (i.e., retronasal olfaction) apparently is similar to what most participants think of as smelling (i.e., orthonasal olfaction), because, as noted earlier, both pathways stimulate the same set of receptors (Pierce and Halpern 1996). What will modify retronasal sensation, however, is the taste and somatosensory information that accompany eating and drinking. Although this makes for a more complex sensation, all olfactory experience, irrespective of orthonasal or retronasal origin, still seems to be characterized by three dimensions: intensity, hedonics, and quality. This observation is based on findings from multidimensional scaling of pairwise similarity ratings of sets of olfactory stimuli (e.g., Schiffman 1974; Schiffman, Robinson, and Erickson 1977; Carrasco and Ridout 1993). Intensity is (relatively speaking) the simplest dimension and appears constrained to the unipolar dimension of barely perceptible through to overwhelming. The hedonic dimension is continuous, but bipolar, with positive and negative poles. The qualitative dimension is by far the most complex and has resisted any attempt at categorization (Lawless 1996). In the main, food odors appear to be characterized by their idiosyncratic resemblance (redolence) to other odors, rather than by some underlying system of organization as evident for color or pitch. Although intensity, hedonics, and quality are psychologically discrete, they clearly interact. For example, departures from optimal (i.e., expected) levels of intensity or quality both result in ultimately greater dislike for a food or drink (Cardello et al. 1985).

Decisions to consume appear to be based on the resemblance of the stimulus to other foods or drinks (i.e., quality/intensity) and whether prior exposure to it had positive or negative outcomes (i.e., hedonics). Continued ingestion appears to be governed, at least in part, by reductions in palatability of the stimulus (i.e., hedonics) over time. In all of these cases the focus of attention is on the overall object (the food or drink in question) and its prior consequences. Focus is not on its specific components nor on other incidental odors that may be present at the time (e.g., the smell of perfume, table polish, etc.). Thus, two key functional aspects of olfaction, in this context, are identifying olfactory food objects and their consequences and discriminating them from the background of other concurrent olfactory stimulation.

Our experience of food and drink odors appears to be a unitary one. That is, foods such as cheese, chocolate, coffee, and wine are not experienced as a series of discrete components but as discrete entities (Livermore and Laing 1998a). This is strongly indicated by laboratory-based research described later on and by our use of object level labels to describe odors (Dravnieks et al. 1986). Not only is treating a food odor as an object logical from a behavioral

Table 6.1   Number of Volatiles Identified in Different Foods and Drinks

Coffee	655	Cocoa	462	Tea	467
Whisky	57	Cognac	68	Milk	49
Bread	296	Rice	100	Potato	140
Fried onion	120	Tomato	400	Carrot	95
Banana	350	Grapefruit	206	Strawberry	360
Ginger	146	Vanilla	190	Pepper	122

Source: Maarse 1991

perspective, the nature of olfactory stimuli themselves may demand it. Nearly all olfactory stimuli that we encounter during ingestion (and more generally) are not pure odorants, but complex mixtures composed of tens or hundreds of volatile chemicals, each contributing in varying degrees to the overall aroma (see table 6.1; compiled from Maarse 1991). Moreover, different brands or varieties of the same food or drink will have diverging combinations and, at least for natural food products, such as vegetables, fruits, and meat, there is likely to be considerable variation in the relative abundance of different volatile components both over longer periods (e.g., as a consequence of drought) and in the short term as well (e.g., as a consequence of senescence). Consequently, the olfactory system has to identify a food that is composed of multiple volatiles that may change over time, against a changing background of other odors. Thus, global resemblance is a more useful heuristic than a careful analysis of features.

Disease Avoidance

Adults go to some lengths to avoid certain types of smell, especially fecal, urinous, and organic decomposition odors (Rozin, Haidt, and McCauley 2000). Such odors engender a markedly negative hedonic reaction that is characterized as the emotion of disgust. Several types of explanation have been advanced as to the origins of this response, many with a psychodynamic bent (e.g., Miller 1997). Nonetheless, there is contemporary agreement that repulsion to such odors is typically acquired during childhood (Peto 1935; Moncrieff 1966; Stevenson and Repacholi 2003) and that the stimuli that most readily engender such responses are those associated with pathogens (Curtis and Biran 2001). In sum, it pays to stay away from decomposing animals and bodily excretions, as these may be potent vectors of disease.

Disgust is clearly important to ingestion, because a foul-smelling food would likely be an indication of microbial contamination (Rozin and Fallon

1987). Likewise, inhalation, or later oral incorporation by touch, could lead to parasitic or microbial infection (Cousens et al. 1996). Such reactions are clearly not monolithic. If they were, then it would be very difficult to deal with our own intimate odors, those of sexual partners, babies, or pets and there is evidence that hedonic responses in exactly these circumstances are modulated by knowledge of the odor's source (e.g., Kalogerakis 1963; Fleming et al. 1993; Stevenson and Repacholi 2005).

The stimuli that evoke these responses are again composed of a vast array of volatiles (e.g., feces; see Pollien et al. 1997), and the same argument that was applied to the identification of odorous food objects applies with equal force here. Object level identification and rapid hedonic reaction should reduce the likelihood of contact, inhalation, and ingestion.

Sex and Attachment

Clearly odor does not play the key role in mate selection and infant attachment that it does in many mammals. However, at a minimum there is some residual function for olfaction in several aspects of human sexuality and in the relationship between kin. At a psychological level, women report that a man's smell is the most important determinant of attraction, and, not surprisingly given this emphasis on smell, women spend something approaching 3.4 billion U.S. dollars per year on scented products (Herz and Cahill 1997). Behavioral evidence is somewhat more equivocal. Although some data suggest that women are attracted to the putative male pheromone androstenone found in male axillary secretions (Cutler, Friedmann, and McCoy 1998; Cowley and Brooksbank 1991; Kirk-Smith and Booth 1980), the research evidence is too weak to make any definitive statements in this regard (see Wysocki and Preti 1998; Doty et al. 1985).

Whether mate choice is actually influenced by smell has been investigated, but only in the most preliminary way. In one closed religious community in the United States (the Hutterites), mate selection does appear to be governed by the choice of partners who have dissimilar human leukocyte antigen (HLA), as in rodents (HLA is equivalent to major histocompatibility complex [MHC] in rats and mice; Ober et al. 1997). These effects appear to be mediated by olfactory cues, at least in rodents (Ehman and Scott 2001), and humans too can discriminate between the smell from rodent urine coming from different MHC strains (Beauchamp et al. 1985). This suggests that HLA-based human olfactory mate selection may occur.

Although at least one failure to obtain evidence of HLA-based mate preference in humans has been reported (Hedrick and Black 1997), people do seem to prefer dissimilar HLA types when smelling sweat (Wedekind et al. 1995). Others have found that HLA similarity is typically higher in couples not achieving pregnancy through in vitro fertilization (Weckstein et al. 1991). In mice at least, and probably also in humans, MHC/HLA is detected by the olfactory system; thus, in this respect odor may contribute to both choice of a mate and ultimately to breeding success.

Another currently controversial issue is whether olfaction can mediate menstrual synchrony in women. McClintock's (1971) study suggested that women who have relatively frequent contact with each other have more synchronized menstrual cycles than women who do not. More recent work, conducted in various countries, has explored the generality of this effect and has been largely supportive of her original finding (e.g., Weller, Weller, and Roizman 1999). Such work has also demonstrated that synchrony results from olfactory cues (Stern and McClintock 1998). Nonetheless, a critical review by Schank (2001) suggests that some of the reported synchrony effects may be artifacts of the recording procedure and that synchrony may not confer any benefits with respect to reproductive success.

Following birth, breastfed infants appear to develop a preference for the smell of the mother's breast pad over that of another lactating mother (Russell 1976). The converse also appears true, that mothers rapidly acquire the ability to discriminate the body odor of their baby from that of other babies of a similar age (Porter, Cernoch, and McLaughlin 1983). Moreover, fathers, grandmothers, and aunts can also recognize the smell of their adult kin (Porter et al. 1986). All of these processes seem to result from mere exposure.

Although in humans many other sensory channels contribute to the formation of a secure attachment between mother and infant, rapidly acquired odor preferences may reflect the vestiges of a process seen at work in other mammals (Sullivan et al. 1991). Because olfaction is a highly emotive sense, the positive affect generated by such smells may be one factor in maintaining attachment.

The olfactory stimuli that are involved in these various processes are, as with the other stimuli already discussed above, typically complex and variable. Thus, for example, the axillary odor of men will vary as a consequence of diet, anxiety, and exercise but will still retain a unique signature for that individual (Kalmus 1955; Doty et al. 1985). In the same manner, the smell of babies' skin, mothers' skin, breast milk, and female sexual secretions are also composed of multiple volatiles, which may vary over time (Nicolaides 1974).

Thus the ability to identify an individual's unique odor, as humans clearly can, must also depend on characteristics of the whole stimulus, the odor object.

Hazard Warning

Apart from disease avoidance, olfactory cues, especially orthonasal ones, can serve to warn of danger. Burning smells are a ready example, but other odorants too can function in this way by virtue of their association with potentially hazardous situations. These include methyl mercaptan, which is added to odorless natural gas and is regarded as the *smell* of gas (Russel et al. 1993). Likewise, the odors of musty hay, onions, and geraniums, signaled gas attacks with, respectively, phosgene, mustard gas, and lewisite, in the trenches of the First World War (Moncrieff 1951). More recently, tobacco smoke has been similarly categorized and is regarded by many as a trigger for asthma and as a toxic and disgusting chemical cocktail (Rozin 1999). As noted in all the other functional cases above, identification at the object level again appears the norm.

Perceptual Expertise

So far the focus has been on the function of olfaction in healthy individuals. There are certain groups, however, who are more dependent on olfaction. These include perfumers and flavorists, organic chemists, sensory evaluation panels, dairy, beer, and wine manufacturers and judges, expert panels (e.g., detecting taints), blind and deaf-blind people, and olfactory scientists. In all cases, these individuals have considerable exposure to odorants and have access to varying degrees of training, which usually involves learning odor names. Because the ability of these groups may tell us to what extent human olfactory ability can be improved and how, a discussion of the relatively limited literature in this area is postponed until later in this chapter.

What Function Implies about Process

The fundamental task faced by the olfactory system involves the identification of a complex mixture of chemicals (the odor object), which may vary over time, against a continuously shifting olfactory background. Behaviorally, meaningful events in the world occur at the level of the odor object. This level is typically equivalent to the object's visual referent and to its verbal label. Things happen as a consequence of either eating the object (e.g., get sick, feel replete) or contemporaneously with smelling it (e.g., discovery of a fire, dinner cooking). Therefore experientially, the object level description of an odor would seem to be the most useful form of representation.

To enable the identification of an odor object, the olfactory system needs to be able to acquire and bind new patterns of stimulation generated by multiple chemical stimuli — new odor objects — consequently olfactory perception should be heavily reliant upon learning. A large number of predictions about the nature of the olfactory system flow from this ecological perspective. (1) Where variation in exposure to odors occurs, either deliberately through manipulation in the laboratory or by natural experiments such as between cultures, one should find differences in olfactory perception for the same stimulus. That is, those familiar with a particular odor object should be better able to discriminate it from other odors and should describe it in a more specific way, relative to those unfamiliar with it. (2) Not only should objective and subjective perceptual characteristics be affected by experience, but so should the affective component of olfactory experience, as this represents the consequences ensuing from previous encounters with it. (3) A heavy reliance on learning must imply a heavy reliance on memory both for odor object recognition and hedonics. Consequently, any damage to olfactory memory — however it is instantiated in the brain — should result in significant deficits in odor discrimination and in appropriate hedonic responses as well. (4) Reliance on learning also implies developmental changes in the ability to discriminate odors as experience is accrued; consequently, developmental changes in olfactory perceptual abilities such as discrimination (improvement over childhood) and olfactory hedonics (increase in variance over childhood) should be expected. (5) Many of the processes that take place in healthy individuals as they acquire odor objects (e.g., predictions 1 and 4) should be accentuated in people with olfactory expertise. Such individuals should be especially good at odor discrimination within their respective domains. The experimental evidence pertinent to all of these points, especially the role of learning, is examined in the next section.

Finally, olfactory experience, in healthy participants, does not occur in isolation. Other sensory modalities, especially taste, somatosensation, vision, and, to a lesser extent, touch and audition, impact both upon how an odor is perceived and whether it is liked or disliked. Likewise, language probably plays an important role in how we respond to odors and possibly in how we perceive them, as the motivational state also does. Therefore any complete view of olfaction, especially in humans, must take into account the often subordinate nature of olfaction relative to language and the other senses and, as in animals, the organism's motivational state. These issues are discussed at the end of the next section.

Is What We Know about the Olfactory System Consistent with This?

Learning

At least two putatively different types of learning process have been identified in human olfaction. The first involves quality and the second hedonics, but whether these need to be viewed as independent systems is not yet fully established. For clarity of exposition we deal with each separately and examine the issue of independence in the hedonic section.

Sensory Systems

Mere exposure to a set of unfamiliar odors results in greater discriminability between the members of that set. Rabin (1988) exposed participants to a set of seven odors, twelve times, after which participants were given a series of paired discrimination trials. On each trial pairs were either composed of two identical odors or two different odors. Participants who had received either of the two control conditions—no preexposure or exposure to a different set of odors—were significantly worse at discrimination relative to the preexposed alone group. A similar finding was made by Jehl, Royet, and Holley (1995), who exposed participants to targets or distractors either not at all, once, twice, or three times in a between-group design. This was followed the next day by paired comparisons between same and different odors pairs, some or all of which had been preexposed during training, depending on the group. Exposure enhanced discrimination. The maximum benefit was obtained following two or three exposures (d' for zero exposure, 1.5; one exposure, 1.8; two exposures, 3.2; and three exposures, 3.8). Although hit rate tended to increase with exposure ($p = 0.07$), most of the effect manifested as a decrease in false alarms.

One concern in these experiments is that participants are required to hold some form of olfactory image during the interval between receiving each member of the pair on test. Although on the surface this would appear to make the studies above less interesting, if anything it suggests the reverse. This is because if the representation becomes more well defined as a consequence of learning the odor object, then the representation should also be enhanced across time improving discrimination on such comparative tasks. Rabin and Cain (1984) demonstrated exactly this relationship. They found that recognition memory performance (a function of discriminability) was significantly related to the judged familiarity of the odors used in testing. Thus greater familiarity predicted better recognition memory performance. Similar results have also been obtained using experimental methods in which familiarity (exposure) is manipulated within the design. Jehl, Royet, and Holley (1997) found that exposure alone, compared with a no-exposure control, enhanced performance on a subsequent recognition memory test.

Rabin (1988) conducted a second experiment that also suggests that familiarity enhances discriminability. The participants' task was, as earlier, to determine whether a pair of odors were same or different. This time, however, the to-be-compared pairs were either composed of two identical stimuli as before or a target followed by a *mixture* composed of the target and a distractor. This experiment used stimuli individually tailored to each participant, so that the familiarity of the target and the familiarity of the distractor were manipulated (familiar vs. unfamiliar). Participants were best able to tell two stimuli apart when both target and distractor were familiar ($A' = 0.9$) and were least able to do so when both target and distractor were unfamiliar ($A' = 0.6$). Yet again familiarity with the stimuli enhanced discriminability presumably via enhanced identification of the two odorants presented in the *mixture* (i.e. both established as odor objects). Conversely, when both target and distractor were unfamiliar, not only would it be harder to identify the target (as it presumably is not yet encoded as an object), but the *mixture* would likely be processed and subsequently encoded as an object itself.

Mere exposure to odors results in other perceptual changes too. It has been known for sometime that certain odors are characterized as smelling of certain tastes, most notably sweetness (Harper, Bate-Smith, and Land 1968). The qualia of sweetness reported by participants when smelling an odor such as cherry, caramel, or vanilla, for example, appears to resemble the qualia of sweetness generated on the tongue by chemicals such as sucrose. Several lines of evidence suggest this (see Stevenson and Boakes 2004, for a more detailed discussion). First, many experimenters have demonstrated that when a sweet-smelling odor is added to a sucrose solution and sampled by mouth, the mixture is reported as sweeter tasting than judgments of the sucrose alone—the sweetness enhancement effect (e.g., Cliff and Noble 1990). Not only has this effect been obtained using category scales, line ratings, and magnitude estimation, it has also been observed using magnitude matching to sucrose (Stevenson 2001b). Second, the effect is odor specific, that is, only certain odorants act to produce the sweetness enhancement effect (Frank and Byram 1988). For example, the degree to which an odor smells sweet is the best predictor of the degree to which that odor will enhance the sweetness of sucrose (Stevenson, Prescott, and Boakes 1999). Third, just as the sweet taste of sucrose acts to suppress the sourness of citric acid, so too do sweet-smelling odors when added to citric acid and sampled by mouth (Stevenson, Prescott, and Boakes 1999). These three characteristics all suggest considerable concordance between the qualia of tasted and smelled sweetness.

One possible explanation of the sweetness enhancement effect is that it results from the way in which participants use rating scales (van der Klaauw and Frank 1996). Clark and Lawless (1994) have argued that when, for example, strawberry odor is added to sucrose solution and participants rate *just* the sweetness of the mixture, they dump similar perceptual characteristics into that rating scale (e.g., strawberryness). Evidence for this has been obtained. When participants who rate just sweetness, as in the example above, are compared with other participants who rate both sweetness and strawberryness, the degree of sweetness enhancement in the latter group is reduced though not eliminated (see Clark and Lawless 1994). This is, however, not a general consequence of adding additional rating scales. If an extra inappropriate scale is added (e.g., meatiness), this has little effect on sweetness enhancement. The question must then be asked: do sweetness enhancement effects provide evidence for the perceptual similarity of sweet tastes and smells?

There are two reasons to think they do. First, the dumping account of the sweetness enhancement effect can not explain why sweet odors act to reduce the perceived sourness of sour tastes, whereas a perceptual similarity account can. Second, dumping provides no basis for understanding *why* certain odor qualities can be dumped into sweetness ratings (e.g., strawberryness) and *why* others cannot (e.g., meatiness). One explanation, which has been supported empirically, is that in the case of odors such as strawberry, vanilla, and cherry, for example, the frequent co-occurrence of sweet tastes with these odors (i.e., flavor) results in such odors acquiring a sweet smell. Consequently participants judge the two qualities of the odor (e.g., strawberryness and sweetness; caramel and sweetness) as perceptually more alike through a process of perceptual learning-acquired equivalence. One result of this is the propensity to show halo dumping, another is that these odors smell sweet.

The central thrust of the argument here concerns the role of learning in odor quality perception. The discussion above of whether odor sweetness is perceptually similar to tasted sweetness was the precursor to a series of experimental reports exploring how certain odors come to smell sweet. Not only do these series of reports support the assumption of perceptual similarity between the sweet characteristics of certain odors and sweet tastes, but more importantly they indicate the potential plasticity of odor quality perception. The first exploration of acquired sweetness utilized a rather cumbersome design, because the exact parameters needed to obtain the effect were not known then. Consequently, Stevenson, Prescott, and Boakes (1995) used a procedure which they thought might emulate the way in which odor sweetness might arise under naturalistic conditions, namely, over many days rather than within a single experimental session.

In this procedure participants rated the characteristics of a series of odors, including two unfamiliar targets, lychee and water chestnut. The key ratings, embedded in a series of other ratings, were how sweet and sour the odor smelled. Participants then returned on a second, third, and fourth day and, on each of these days, sampled a series of solutions, some of which contained one odor consistently dissolved in sucrose and the other consistently dissolved in citric acid solution. Trials were disguised as triangle tests and participants were told that they were participating in an experiment examining olfactory psychophysics. On the fifth day participants returned and smelled all the odors sampled on day one, rating each again by using the same set of scales. For the sweet-paired odor, there was a significant increase in odor sweetness and a significant decrease in odor sourness (fig. 6.1, top). For the sour-paired odor, the reverse obtained, it smelled more sour and less sweet, postconditioning (fig. 6.1, bottom). Participants, when questioned at the end of the experiment, were unable to identify the actual aim of the experiment.

These basic effects of acquired sweetness and sourness have now been obtained under a variety of different conditions; both within a single session and with as few as three pairings, indicating that this form of learning may occur rapidly (see Stevenson, Boakes, and Prescott 1998; Prescott 1999; Stevenson and Case 2003). Not only do odors that have been paired with sucrose get to smell sweeter, they also act to enhance the sweetness of a sucrose solution to a greater extent than they did prior to conditioning and relative to control odors (Prescott 1999). In all of these studies, considerable attention has been

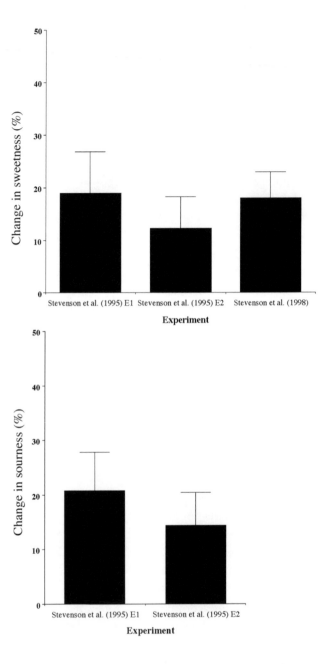

Fig. 6.1. (*Top*) Mean change (plus standard error) in odor sweetness (post- minus pre-test) from experiments 1 (E1) and 2 (E2) (Stevenson et al. 1995, 1998). (*Bottom*) Mean change (plus standard error) in odor sourness (post- minus pretest) from experiments 1 and 2 (Stevenson et al. 1995).

given to the issue of participants' awareness of the experimental contingencies. Using several measures, such as postconditioning tests of awareness, we have not obtained any reliable evidence that knowledge of the procedure relates to obtaining increases in perceived odor sweetness or sourness (Stevenson, Boakes, and Prescott 1998; and for critical discussion, Lovibond and Shanks 2002; Stevenson and Boakes 2004).

Odor-taste learning has not only been demonstrated to be rapid and to occur with minimal awareness, but it has also been shown to occur with tastes other than sucrose and citric acid, notably bitter (sucrose octa-acetate and quinine) and for mixtures of salt and umami. Thus the effects are not specific to sweet and sour tastes but occur more generally (Yeomans et al., submitted). Moreover, we have recently demonstrated that odors can acquire trigeminal-like qualities too (Stevenson and Case, submitted). Three odors were used, one of which was sniffed in combination with menthol, another one in combination with acetic acid, and the final one on its own as an exposure control. On test, participants rated the menthol-paired odor as smelling cooler and less warming than the control odor. A further posttest revealed that when the menthol-paired odor was added to a dilute solution of menthol, this mixture was judged to smell more cooling than when the control odor was added. Similarly, the acid-paired odor, when added to a dilute solution of acetic acid, was judged to smell more pungent than when the control odor was added. Finally, we have just completed a further study in which we explored whether odors could acquire fatlike properties (Sundqvist and Stevenson, in preparation). Following pairings with either low-fat or high-fat milks, an odor paired with high-fat milk was found to enhance fattiness ratings of a midrange fat content milk, more than an odor paired with a low-fat milk.

All these findings mean that odor quality can change as a result of experience. However, it could reasonably be argued that the acquisition of tastelike, fatlike, or trigeminal-like properties is a special consequence of the intimate relationship between these senses during ingestion and routine smelling. Nonetheless, there are now solid grounds upon which to argue that the plasticity of odor perception evidenced by these studies is a general phenomenon, as discussed next.

Two related sets of findings suggest that odor quality perception may change under conditions in which *only* olfactory stimuli are present. The first are laboratory-based studies of odor quality acquisition. In these experiments participants are exposed to a purely olfactory stimulus, composed of a mixture of two odors. As described below, this procedure can affect the reported quality of the mixture elements postconditioning, their judged similarity to each other, and their discriminability. The second line of evidence comes from naturalistic studies of perceptual learning, in which groups of participants presumed to differ in olfactory expertise of one form or another (though typically of wine) demonstrate greater ability to distinguish between odors relevant to their area of expertise than naïve controls. Yet again, note the basic resemblance of all the processes so far discussed in this section—exposure leads to a detectable change in some property of odor perception, as measured by odor quality, similarity, or discriminability.

Under certain conditions, mere exposure to a mixture of two odors can result in the elements of that mixture coming to acquire properties from each other, smelling more alike and being less discriminable (Stevenson 2001a, 2001c, 2001d). The early experimental investigation of these effects followed from the studies of odor-taste learning and so initially focused on the perception of odor quality. In these experiments, participants smelled a series of odors (four, termed here A, B, X, and Y) and rated them for their qualities. That is, odor A was rated for the degree to which it smelled A-like, B-like, X-like, and Y-like (e.g., if A was citral, then it would be how lemonlike does this odor smell?). These odors were then combined into two sets of mixtures (either AX and BY or AY and BX) and participants were randomly allocated to one of these sets. Each member of the set was then smelled several times. This was followed by a subsequent experimental session one week later in which the four odors alone (A, B, X, and Y) were rated again, using the same scales as in the pretest.

Three findings emerged over a series of four experiments (Stevenson 2001a, 2001c). First, changes in odor quality do occur. (1) Mixtures of L-carvone (or cis-3-hexanol) and p-anisaldehyde (musty smelling) resulted in both of the former smelling mustier. (2) Mixtures of terpineol (or p-anisaldehyde) and cis-3-hexanol (green smelling) resulted in both of the former smelling greener. (3) Mixtures of methyl salicylate (or guaiacol, or champignol) and cherry, resulted in all of the former smelling more cherry-like. (4) Mixtures of cherry (or citral) and guaiacol (smoky smelling) resulted in both of the former smelling smokier. (5) Mixture of champignol and citral (lemon smelling) resulted in the former smelling more lemon-like.

One problem with the procedure above is that of scale definition. Although most participants have a clear idea of what sweet means, their concept of the odorlike qualities measured above is likely to be considerably more varied. Consequently, a second dependent variable was also obtained in these studies in a further posttest similarity. This dependent variable allows the capture of changes in quality without reliance on specifying exactly what they may be, thus avoiding the problem of participants' idiosyncratic use of the odor quality scales. The similarity measure turned out to be considerably more reliable. Whenever a change in odor quality was obtained, as described in the preceding paragraph, consistent changes in odor similarity were typically obtained as well. These measures were also found to correlate, such that larger changes in reported quality went along with greater increases in judged similarity.

The second finding to emerge was that not all mixture pairs demonstrate this effect. Following considerable unpublished pilot work, the data pointed to at least two possible reasons why this might be so. First, when the odors are both unfamiliar changes do not seem to occur and optimum conditions pertain when one odor is familiar and the other unfamiliar. Second, participants who are able (on a subsequent posttest) to identify one or both components in the mixture, evidence greater acquisition. In this case, one possible explanation concerns scale definition, in that learning effects are most likely to be detected in participants who share the experimenter's label for the odor. Thus, if they identify the component using that label, they probably share his definition. Notwithstanding this partial explanation, the parameters that govern odor-odor learning have not been adequately explored and a systematic investigation might tell us more about the processes underlying this form of learning.

The third finding to arise from the study of odor-odor learning concerns the behavioral consequences of the conditioning procedure. The original evidence of acquired odor qualities and similarity rested on self-report. Two experiments have been conducted which demonstrate that mixture pairing also results in reduced discriminability (Stevenson 2000c; Case, Stevenson, and Dempsey 2004). The first presented participants with two odor mixtures (e.g., AX and BY), followed by a triangle-discrimination procedure (or oddity test), composed of trials in which the paired odors were discriminated (e.g., A vs. X vs. X) and trials where the nonpaired, but equally exposed odors were discriminated (e.g., A vs. Y vs. Y). Discrimination was significantly poorer when odors had been previously experienced in a mixture relative to those equally exposed but not experienced as a mixture. Although these findings *imply* reduced discriminability, this cannot be strongly inferred from the design, especially as it is known (see above) that exposure alone enhances discriminability (Rabin 1988; Jehl, Royet, and Holley 1995). That is, the effect reported in this experiment could have been obtained by changes in the nonpaired but exposed stimuli (from the example above, the A vs. Y or B vs. X comparisons).

As a result we ran a second study (Case, Stevenson, and Dempsey 2004) which compared three conditions, two odors presented alone (A, X) as often as the two other odors experienced in a mixture (BY) and a further two odors presented *only* during the discrimination phase (C, Y). If experience as a mixture reduced discriminability as expected, then discrimination should be reduced for the B vs. Y pair relative to both the exposed (A vs. X) and, crucially, the nonexposed controls (C vs. Y). This was exactly what was observed, suggesting that pairing two odors in a mixture can result in the pair being judged as less discriminable than controls. A further feature of this experiment was that it also included similarity ratings of the three odor pairs. These yielded an identical pattern of findings to the discrimination data and, not surprisingly, were correlated with them too.

A second line of evidence suggesting that odor perception is affected by learning is derived from the study of perceptual learning, which more often than not has explored differences accrued outside of the laboratory. Two issues are of particular interest here. First is the discernment of whether the effect of exposure alone enhances perceptual expertise, as this is the most direct corollary of the experimental findings above. Second is the examination of the role of language, especially in experts. Clearly the training of experts involves the acquisition of *both* experience with odors and the acquisition of appropriate labels for those experiences. Label learning can increase discriminability above that of exposure alone. Rabin (1988), in addition to the experimental findings described above, also ran a further experimental group who learned labels for the exposed stimuli. On testing, this group was significantly better at discrimination relative to the exposure-alone group. This type of improvement could result from several mechanisms, which can be divided into those that involve some alteration in the percept (i.e., sharpening the target, ignoring the background) and those that use the verbal label as a means of assisting retention, both in the short term across serially presented discrimination trials or longer term, over days or weeks (see Lyman and McDaniel 1986). Not only do labels benefit experts by enhancing discriminability, irrespective of mechanism, they also confer a further advantage in facilitating cross-talk between perceptual and semantic level representations. That is experts can better match their and other experts descriptions of wine, to the target, than novices (see Lawless 1984; Solomon 1990).

The human olfactory perceptual learning literature is not well developed. Most focus has been given to expertise with respect to wine, which although involving other senses, clearly has a significant olfactory component. Unfortunately, few studies have directly compared naïve participants (hereafter, perceptual novices) with those who regularly drink wine but who have no formal training (hereafter, perceptual experts) against those who have both perceptual expertise and formal training in wine terminology (hereafter, semantic experts). A study of central importance, which has made such comparisons, was reported by Melcher and Schooler (1996), who examined whether discriminative ability differed between these three types of groups; namely, participants who never drank wine, those who drank it regularly but with no formal training, and those who both drank it regularly and who had had formal training in wine terminology.

Melcher and Schooler's (1996) study comprised three phases. In the first, the exposure phase, participants sampled one wine. In the second phase, all participants, irrespective of expertise, were randomly allocated to either a condition in which they completed crossword puzzles or to a condition in which they were asked to write a description of the wine sampled in the first phase. In the third phase, completed by all participants, four wines were presented, one of which was the target encountered during the exposure phase. For each wine, participants were asked to rate how certain they were that this wine was the target encountered during the exposure phase. To clarify the findings here, the results are reported separately for the crossword puzzle group first, then for the wine description group second.

Novice participants, that is, those who rarely drank wine and thus had little perceptual expertise, performed at chance level on the discrimination task (see fig. 6.2). Participants with perceptual expertise but no formal wine training performed significantly better and above chance, as did the group who had both perceptual and semantic expertise. These latter two groups did not differ significantly. Thus, consistent with the general thrust of this section, exposure alone significantly improved discrimination.

The participants who completed the wine description evidenced a different pattern of findings (see fig. 6.2). The semantic experts who had both perceptual experience and formal wine training performed best, at a level akin to that in the crossword puzzle condition described above. However, attempting to produce a written description exerted a markedly negative effect on discrimination in the group who just had perceptual expertise. Their performance was worse than the semantic experts and *significantly lower* than their group's performance in the crossword condition. Thus attempting to shift from a semantic level description to a perceptual one on the test adversely affected their performance. The novice group actually performed slightly better and above chance here, as the description afforded them some clue during the discrimination phase. In sum, these data suggest that exper-

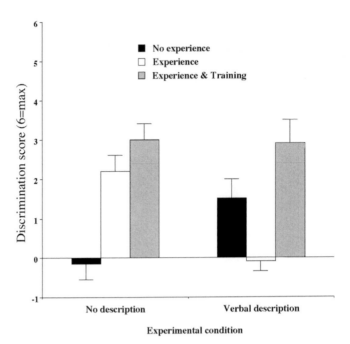

Fig. 6.2. Mean discriminative performance (standard error) of participants with no experience of wine, experience, and experience with formal training (data adapted from Melcher and Schooler 1996). Participants in the verbal description condition were asked to write a description of the wine prior to the discrimination test.

tise has two dissociable components, a perceptual one based on exposure and a semantic one based on formal training linking terminology to sensory experience.

A slightly different approach to the role of perceptual expertise has been taken by Walk (1966) and Owen and Machamer (1979), who examined the effect of exposure and training on wine discrimination in perceptual novices. Walk (1966) found that exposure enhanced discrimination, primarily through a reduction in false alarms. Training (feedback and/or numeric labels) was no more efficient at producing this effect than exposure alone. Owen and Machamer (1979) also found that discrimination could be enhanced following exposure to wine in novices. However, in their case, performance change manifested as both a reduction in false alarms and an increased hit rate. A similar approach has also been adopted in a study using beer rather than wine. Peron and Allen (1988) compared various types of training (exposure, terminology, both) on participants who were initially poor at discriminating different types of beer. They too found that relative to nonexposed controls, only exposure enhanced performance on both a triangle test of discrimination and a similarity test for identical stimuli.

The examples above are probably representative for whole classes of olfactory stimuli; that is, the mechanisms utilized for wine and beer probably extend to perfume and other odorous products, where the expertise is not likely to generalize far from the specific stimuli encountered within that set. One case where this is unlikely to be true is in participants who have lost their sight (and in deaf-blind people too). Such individuals arguably depend more on their sense of smell (see James [1890] for his description of Laura Bridgman and Julia Brace and Williams [1922] for a description of the deaf-blind Willetta Higgins). If this is correct, then although not differing in terms of exposure relative to sighted participants or in innate ability, they should pay more attention to the olfactory domain. Several studies confirm this, in that blind participants are significantly better at labeling a wide variety of familiar odors relative to sighted controls (e.g., Rosenbluth, Grossman, and Kaitz 2000; Murphy and Cain 1986). Blind participants effectively have an enhanced terminology with which to describe odors that we all routinely encounter. That is, they appear to have a selective semantic expertise for day-to-day odors.

In sum, studies of olfactory perceptual learning conform to the general pattern of observations described above, namely that mere exposure affects odor perception through the passive acquisition and storage of the odor percept. This may result in enhanced discriminability between separately exposed items or in reduced discriminability for items exposed together in a mixture and encoded as an object. Such effects can occur rapidly and take place with minimal conscious awareness. More importantly, they provide a strong body of evidence that odor quality perception and its behavioral corollary discrimination are markedly affected by experience.

Hedonic Systems

So far we have examined the effects of experience as they apply to odor quality. Changes in hedonic responses to odors can occur under some of the same conditions examined above, notably following mere exposure and after pairing with tastes. Not only are these observations important because of the plasticity that they imply for odor hedonics, but also because they can be used to explore whether olfactory hedonics and quality are in fact dissociable psychological systems. In addition to exposure- and taste-based learning, hedonic responses to odors can also be modified by other associative mechanisms, notably those involving a delay between the odor and some subsequent event. These primarily include calorie-based learning and conditioned taste aversions. These are examined in the latter part of this section.

Not only can mere exposure enhance the discriminability of odors, it also appears to affect the liking for them too. Several studies have confirmed that greater familiarity strongly correlates with increased liking (Engen and Ross 1973; Lawless and Cain 1975; Rabin and Cain 1989). Clearly, such studies must depend on the selection of odors, because some familiar odors will be judged as highly unpleasant (e.g., sweat, garbage). Whether this relationship reflects the effects of sensory exposure or the ability to identify odors *as a consequence* of enhanced discriminability produced by exposure is an important question, because it points to either the similarity or difference of this effect to the perceptual exposure results discussed above (cf. Rabin 1988). Familiarity is logically correlated with nameability (e.g., Stevenson 2001c), and because nameability markedly affects liking (see Herz and Von Clef 2001 [isovaleric acid can be labeled positively as Parmesan cheese or negatively as vomit, with predictable effects on hedonic evaluations] and Ayabe-Kanamura, Kikuchi, and Saito 1997), the question remains whether the relationship between familiarity and liking is mediated by semantic knowledge about the odor, rather than simply being a consequence of exposure.

Experimental studies of exposure, and its effect on liking suggest that mere exposure to a pure odorant results in drifts toward hedonic neutrality rather than toward liking. Cain and Johnson (1978) assessed the liking of a group of participants for a range of odors, including four targets. Some participants were then exposed to one target 55 times, with the target being pleasant, neutral, or unpleasant; the remaining participants received no exposure at all. Participants then rated liking for all the odors again, including the targets. For participants exposed to the pleasant target, liking ratings *decreased*, whereas for participants exposed to the unpleasant target, liking ratings increased. Thus, massed exposure tended to result in indifference, rather than an overall trend to increase liking.

Arguably, 55 exposures in a relatively short time may not be representative of the way in which mere exposure works under naturalistic conditions. In this respect, several studies have found that exposures to foods, where the olfactory component is a significant contributor to the overall sensory experience, *do show* increases in pleasantness when exposures are more limited (i.e., <10) and not massed (e.g., Pliner 1982, Birch and Marlin 1982, Crandall

1984). In most cases exposure does appear to enhance liking purely through sensory means, because the stimuli were readily identifiable (e.g., doughnuts in the Crandall [1984] study) and identifiability thus remained constant across the experiment. It seems then that both sensory experience with an odor and ability to name it when the stimuli are masked, may both contribute to changes in liking consequent upon exposure.

Procedures that involve pairing an odor with a taste also result in hedonic changes. Zellner et al. (1983) presented participants with unusual flavored teas, either with or without sugar. They found that on a posttest measuring liking, ratings were most enhanced for the sweet-paired tea relative to the others. Not only can pairing with a taste enhance liking, it can, with an unpleasant taste, decrease it. Baeyens et al. (1990) found that a fruit odor paired with the bitter taste of polysorbate Tween-80 was liked less than a water-paired control. As Rozin, Wrzesniewski, and Byrnes (1998) also observed, obtaining positive hedonic conditioning with sucrose was more difficult; no significant effect was obtained by Baeyens et al. (1990). Recent work by Yeomans et al. (submitted) suggests that this may be because of the far greater variability in liking for sweet tastes, relative to the high level of agreement for disliking bitter or sour tastes. Consistent with this is the observation that hedonic changes in odor liking following pairings with unpleasant taste have been replicated many times (see De Houwer, Thomas, and Baeyens 2001), whereas hedonic changes have been far more difficult to obtain when using sucrose (see Stevenson and Boakes 2004).

An important feature of these studies is their methodological and conceptual similarity to the odor-taste acquisition studies discussed earlier. The principle difference is simply the dependent variable, which in the odor-taste studies is sweetness (or other taste-related descriptor) and in the hedonic studies liking. An important and fascinating question is whether these perceptual and hedonic changes represent separate and thus dissociable learning systems. Before reviewing the evidence, it is worthwhile reflecting on how these two phenomena *might* be related. The first possibility is that they are totally dissociable, that is, changes in liking could occur independent of any change in say odor sweetness or bitterness. The second is that changes in an odor's perceptual properties *result* in changes in liking (the converse is equally plausible). The third is that when liking sweetness or bitterness is measured, the measures are synonymous, that is there is only one underlying process and participants simply rate a negative hedonic change as bitter and a positive hedonic change as sweet.

The single most important feature of odor-taste learning is its marked resistance to interference (more below). Similar findings have been obtained when the dependent variable is changed in liking. Baeyens et al. (1995) found that after six pairings of an odor and taste, two blocks of presentation of the odor alone (four times on each block) did not eliminate the acquired hedonic change. Baeyens, Crombez, et al. (1996) also made a similar finding. No direct attempt has yet been made to see whether these hedonic changes are also resistant to counterconditioning, as odor-taste learning is. However, data from experiment 4 of Baeyens, Crombez, et al. (1996) suggest that counterconditioning may occur. On some trials the target odor was paired with Tween-80. This pairing was always preceded by another odor. On other trials the target was paired with sucrose and this pairing was always preceded by water. When the target was evaluated, participants who liked sucrose now liked the target; that is, the positive hedonic value of sucrose appeared to countercondition the negative hedonic response induced by Tween-80. For participants who did not like sucrose, they demonstrated a dislike for the target odor. Although not a counterconditioning experiment, these results do suggest that hedonic changes *may* be susceptible to counterconditioning. This would be consistent with findings in the hedonic conditioning literature for stimuli other than odors, which show that counterconditioning, but not extinction, can occur (see Baeyens et al. 1989). In sum, the interference data suggest that changes in the taste properties of odors are resistant to extinction and counterconditioning, whereas hedonic changes are resistant to extinction, but not to counterconditioning.

More compelling and as yet unpublished data have recently emerged from Yeoman's laboratory. He found that following odor-taste pairings, a caloric preload affected participants' hedonic judgments for the odor but had no impact on ratings of odor sweetness. This finding provides the strongest evidence to date that changes in odor sweetness are dissociable from changes in odor liking.

Additional support for this conclusion comes from two further sources. First, as described earlier, multidimensional scaling of odor similarity ratings suggests three dimensions: quality, hedonics, and intensity (e.g., Schiffman 1974; Schiffman, Robinson, and Erickson 1977; Carrasco and Ridout 1993). If quality and hedonics were synonymous then only two dimensions would be expected. Second, odors that are equated for liking but differ in quality can be discriminated (e.g., Rabin 1988). Clearly, this suggests that odor hedonics and quality cannot be synonymous at least under these conditions. Taken together, the available evidence points then to the first possibility, that odor hedonics and quality represent separate learning systems.

Changes in liking for an odor can also be obtained under conditions where the odor stimulus is followed at some later point by a consequence. Thus, there is both an appreciable delay of time between the two events and both events are readily discriminable. This is in contrast to all of the examples that we have reported so far, where learning occurs under conditions of simultaneous presentation of often hard-to-distinguish stimuli (i.e., an odor and a taste). The most well-studied examples are conditioned taste aversions, which are rapidly formed and are relatively easy to extinguish *if* the target food/drink is ingested again (Schafe and Bernstein 1996). Surveys reveal that these occur commonly (in about 50% of college populations; Midkiff and Bernstein 1985) and that they can be readily obtained by pairing a preferably novel food with any source of nausea (e.g., Bernstein and Webster 1980). Positive affect can also be conditioned by using such trace procedures. Several studies in humans have shown that food that produces repleteness can become associated with its odor, producing an increase in liking for that food's smell (see Capaldi 1996, for review). Finally, odors can become associated with more proximate good or bad events. Baeyens, Wrzesniewski, et al. (1996) found that an odor placed in an office toilet was preferred over a control odor, but only in participants who enjoyed going to the toilet.

Experience then can affect both the perceptual and hedonic characteristics of odors. The key conclusion here is that olfactory perception and hedonics are highly plastic. If this is correct, then it would suggest that considerable variation in both liking and quality for the same odors should be observed between cultures that differ in their history of olfactory exposure.

Culture

Considerable variation is evident in odor preferences between cultures. Although North Americans regard wintergreen favorably, British participants do not (Moncrieff 1966). South East Asians enjoy the pungent onionlike/fruity odor of the Durian and the smell of fermented fish sauce, both of which are disliked by many Europeans (Pangborn 1975). Conversely, Europeans enjoy the off-flavor of blue cheese, the vomitous odor of Parmesan, and the rotting smell of hung meat (Jones 2000). Tibetans enjoy tea mixed with yak butter (Moore 1970), whereas the pastoralist Dassanetch of Ethiopia consider all smells associated with cattle to be good; consequently hands are washed in cow urine, men cover their bodies in manure, and young women enhance their attractiveness by smearing clarified butter onto their bodies (Classen

1992). Breath odors are generally avoided in the West, yet Arabs regard it as shameful to avoid smelling each other's breath. On the basis of these anecdotes one might reasonably expect that hedonic reactions to odors, especially ones that are not routinely encountered within a culture, should differ and this has been confirmed in several studies (e.g., Davis and Pangborn 1985; Schaal et al. 1997; Wysocki, Pierce, and Gilbert 1991). However, more intriguing is the question of whether a culturally novel odor is perceived in different ways too, as one might expect from the learning data presented above.

Ayabe-Kanamura et al. (1998) had German and Japanese participants smell a range of culturally specific (e.g., marzipan for Germans, roasted tea for Japanese) and international odors (e.g., peanuts, coffee). As expected, liking ratings for culturally specific odors were higher in the group that routinely encountered that odor, but some differences, albeit smaller ones, were also evident for the international set. The qualitative aspects of the odors varied in a similar way. Whereas German participants judged fermented soy beans as smelling like cheesy smelly feet, dried fish of excrement, soy sauce of fresh bread, and roasted tea as fishy, Japanese participants judged them differently, in line with their specific use as food products. In the same manner Japanese participants judged marzipan as smelling oil-like or like sawdust or bees wax and Pernod as disinfectant-like odors that were not judged so by German participants.

A further study conducted by Ueno (1993) examined whether perceptions of 20 Japanese food odors differed between Japanese and Sharpe (Nepalese) participants. Judgments were made by participants, arranging the odors into groups based on their similarity. These ratings were determined by cluster analysis, which revealed that fishlike odors were evaluated in a different manner by Sharpe participants, who presumably rarely encountered such smells. Although, for the Sharpes, fish odors did not exist as a discrete cluster, they did for Japanese participants. Although these two studies imply the sort of perceptual differences that would be expected through a differential exposure history, the paucity of research in this area and the absence of studies using discrimination makes a final judgment of the role of culture on odor quality perception difficult. Nonetheless, the two studies described above are highly suggestive.

Development

If the olfactory system is heavily reliant on learning, then two developmental predictions flow from this proposition. The first prediction is that the ability to discriminate odors should improve from infancy to adulthood as a direct consequence of the progressive amassing of olfactory experience. Second, initially, in infancy, hedonic reactions to odors should be considerably less marked than in adults. This is because the experience necessary to produce a hedonic response has either not yet taken place or the infant is not yet sufficiently developed to be able to *understand* why something like feces or vomit, for example, are bad. As described below, there is support for both propositions.

Processes that strongly correlate with discriminative ability (see De Wink and Cain 1994), such as recognition and odor-naming tasks, do show the predicted pattern of improvement with age. Russell et al. (1993) used data obtained from the National Geographic Smell survey. For all six odors used, eugenol, isoamyl acetate, methyl mercaptan, rose, androstenone, and galaxolide, identification of the most appropriate label was consistently poorer in the youngest sample (age, 10–19; $n = 80,533$) compared with young adults (age, 20–29; $n = 152,886$). Participant responses were also subjected to multidimensional scaling, which revealed two interesting findings. First, that odor space (i.e., how the odors were qualitatively related to each other) remained stable between 30 and 60, thus participants younger than this (and older) perceived the odors in apparently different ways. Second, and relatedly, the difference in the youngest participants (age, 10–19) appeared to be in the hedonic dimension, in that there was relatively little separation between methyl mercaptan (gas) and isoamyl acetate (banana) when compared with all older groups. Relatedly, poorer naming ability in younger participants, relative to adults, have been obtained by Doty et al. (1984) using the University of Pennsylvania Smell Identification Text (UPSIT), and by Lehrner et al. (1999) and Cain et al. (1995), using familiar odors. In sum, as Lehrner and Walla (2002) concluded in a recent review article, odor-naming ability is poorer in children than in adults. The suggestion here is that at least some of this difference results from failures of discrimination.

The suspicion that discrimination may account for poor odor naming is strengthened by studies of odor memory. Cain et al. (1995) conducted a further experiment (experiment 3) in which they assessed the speed with which younger children (mean age = 6), older children (mean age = 10), and adults (mean age = 30) could acquire odor names to familiar and unfamiliar odors. Both self-generated names and applied labels were used, but these did not differ by age or odor type. The younger children had an initially lower baseline for both familiar and unfamiliar odors, but soon acquired the names for the familiar odors to a level equivalent to that of the older groups, suggesting

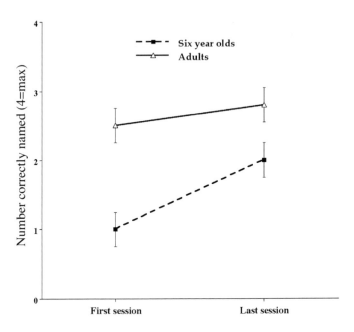

Fig. 6.3. Mean correctly named (standard error) on the first and last session for a task in which children and adults learned vertical labels for four unfamiliar odors (data adapted from Cain et al. 1995).

they were as able at learning odor names as adults. Their performance was markedly worse for the unfamiliar odors relative to the other age groups (see fig. 6.3). This can be interpreted as a failure to discriminate between these odors, which would have proved far less familiar to this group than to the older participants—a conclusion also reached by Cain et al. (1995).

This position is further supported by studies of odor recognition memory. Hvastja and Zanuttini (1989) examined odor-picture-paired associate learning in 6, 8, and 10 year olds. They included both an immediate and delayed recognition memory test of the target set. Immediate performance was worse in the youngest children. For delayed recognition at one month, performance decreased most of all for the older children, but remained stable for the youngest, suggesting a greater reliance in this group upon perceptual encoding rather than on both semantic and perceptual encoding in the older group (see also Lehrner et al. [1999] for a similar conclusion). Jehl and Murphy (1999) found that recognition memory was significantly worse for 7–10 year olds following a 20-min interval than for 11–15 year olds. Likewise, Larjola and Von Wright also found poorer recognition memory in 5 year olds (relative to 10 and 15 year olds) for a set of odors, both on immediate test and at a delayed test one month later. As before, performance was worse in the older age groups on the delayed test relative to the group's performance on the immediate test, although the youngest group was still poorest of all.

Direct tests of discriminative ability in children have only recently been completed. Stevenson, Sundqvist, and Mahmut (submitted) explored olfactory discrimination in 6 year olds, 12 year olds, and adults. Using a set of unfamiliar odors, and the triangle test of discriminability, they found that 6 year olds were significantly poorer at discrimination (mean = 44% correct) than 12 year olds (mean = 62% correct) and adults (mean = 68% correct). The latter age groups did not differ. This age-related difference was obtained under conditions in which naming probably played no effective role, as the odors were all hard to name and an articulatory suppression task was employed (repeating out loud "the, the, the . . ."). These effects were not a consequence of task comprehension, because the youngest children performed at a far-superior level on a visual discrimination task. In a second series of studies (Stevenson, Mahmut, and Sunqvist, in preparation), the same age-related differences were obtained with a familiar set of odors (and replicated under conditions of articulatory suppression) and with a different moderately familiar set of odors too. The effect size of the age-related difference was significantly larger in the latter case. Yet again, a visual control condition revealed similar levels of performance between the 6 year olds and the adults, ruling out task difficulty as an explanation. Overall, both studies of naming, name acquisition, recognition memory and discrimination, suggest poorer performance in younger children, consistent with less exposure to odors and thus a more impoverished odor object memory store.

That there are major developmental changes in hedonic responsiveness to odors is widely agreed (see Engen 1982; Moncrieff 1966; Rozin, Haidt, and McCauley 2000). At birth, and *unlike* with taste stimuli (see Steiner et al. 2001), there is little difference in facial or autonomic responses to odors that adults find pleasant (vanillin) or unpleasant (butyric acid; Soussignan et al. 1997). With somewhat older children (<5 years), Peto (1935) found that they evidenced little emotional reaction to anise, perfume, lemon, camphor, carbol, and putrid or fecal smelling odors. This progressively changed as older groups were examined. Engen (1974) examined odor preferences in 4 and 7 year olds and adults for a range of odors including heptanol (sickening, oily, moldy) and safrole (licorice, sweet, aromatic). Although the rank order of preferences for the younger and older groups were similar, the crucial difference was the hedonic range, which was considerably smaller in the 4 year olds (0.4), relative to the 7 year olds (0.9) and adults (2.0). That is responses became more polarized with age. Evidence of appropriate hedonic responses to odors that adults dislike has been observed in 3 year olds, by asking them to point to one Sesame street character if they liked the odor (Big bird) and another if they disliked it (Oscar the grouch). As with Engen's (1974) findings, rank-order preferences were substantially similar to adults, but the range was again more restricted in children.

Developmental changes in liking have also been observed in somewhat older children. Stevenson and Repacholi (2003) found that 8 year olds were relatively indifferent to the smell of acrid male sweat, but that 17 year olds strongly disliked it. This response to a large extent depended on identification in the older age group, but cueing the nature of the stimuli (by indicating that all the smells in the study came from people's bodies) had a large effect on the 17 year olds, but no effect on the 8 year olds. A similar observation was made for the other odors used in the study. Identification tended to move hedonic responses in more extreme directions, such that it led to caramel being rated as more pleasant and androstenone (identified as sweat/urine/dirty clothes) as less pleasant. As with the other studies reviewed above, the general observation is that with increasing age, more adultlike hedonic responses emerge, a probable consequence of being able to learn that some odors are indicative of bad or contaminating substances (see Rozin, Fallon, and Augustoni-Ziskind 1985). Unless you know what sweat or feces *mean*, why should they smell bad?

Neuropsychology

To the extent that object recognition is essential for routine odor perception and that such objects are acquired and stored in memory, then damage to this odor memory system should result in the following deficit. Odors should be detected (i.e., present/absent, degree of presence), but they should have no quality, that is, equating for intensity, two odors such as cheese and cut grass, for example, would be indistinguishable (i.e., a qualitative discrimination deficit). Retrospectively, as nobody has as yet *specifically* searched for this deficit, it should be manifest in the extant literature in at least two ways. In the first way, participants are able to detect odors (e.g., threshold sensitivity) but unable to correctly name them or recognize them on a memory test—both these tasks being reliant upon successful qualitative discrimination. Such findings would be consistent with the predicted deficit but would not be conclusive, because (a) odor naming/memory can be affected by a variety of olfactory problems other than discriminability (see Doty 1997) and (b) poor performance could be evidence for generalized cognitive deficits, such as, for example, the capacity to attend.

The second way in which the deficit could be identified is by looking for cases where odor detection is intact, but odor quality discrimination is not intact. This time concern would have to be raised about the serial nature of olfactory discrimination tests, in that periods of time typically elapse between presentations, in which case it may be that participants cannot retain a representation of the stimulus over time. Here, adequate control testing using either odor intensity discrimination and/or discrimination testing in another modality would argue against this conclusion. The body of evidence reviewed below offers support for the suggested dissociation between detection and discrimination (i.e., between the intensive and qualitative dimensions of olfactory experience). Unfortunately, as these studies are motivated by varying rationales, different from the one being examined here, the majority of findings are of the first sort and relatively few are of the second. In addition, certain studies utilize patients with unilateral damage; and, for sake of clarity, we focus here primarily on the general nature of their deficits, rather than on the extent to which olfactory functioning is lateralized.

Odor identification declines quite dramatically during normal ageing (see fig. 6.4; Doty et al. 1984). It is also apparent that odor detection thresholds increase with age too (e.g., Cain and Gent 1991; Kareken et al. 2003; Stevens and Dadarwala 1993). Across studies, thresholds for the elderly are typically between 2 and 15 times higher for most odors (Schiffman 1993). To what extent then can declines in identification be accounted for by changes in detectability? Several studies suggest that detectability does not play a major role. First, Cain et al. (1995) examined both threshold and naming ability in 100 adults aged from 18 to 90 (data pertinent to children was reported earlier). Regressing age, the Boston Naming Test (naming pictures), and threshold level against unprompted odor identification score, revealed independent effects for age and the Boston Naming Test, *but not for threshold.* That is age alone, independent of threshold, predicted some unique variance in odor identification ability. Second, Schiffman (1992) tested 143 participants aged between 10 and 80. Participants' ability to name odors, among other things (more below) declined with age. However, although thresholds for three tested odors were higher in older participants, this did not significantly differ between age groups (and see Larsson and Backman 1997, for a similar result). Third, Stevens, Cain, and Demarque (1990) examined odor recogni-

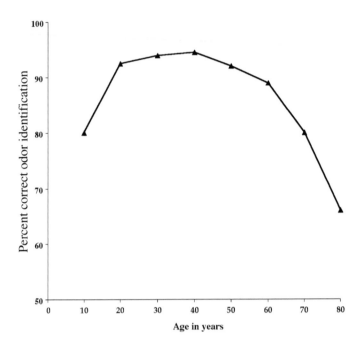

Fig. 6.4. Mean percent correct odor identification by age on the University of Pennsylvania Smell Identification Test (data adapted from Doty et al. 1984).

tion memory in the young (mean age = 22) and elderly (mean age = 73) and used odors at three levels of suprathreshold intensity (weak, medium, and strong). If alterations in threshold were important, then recognition memory performance should improve with increased stimulus intensity (i.e., compensation). Recognition memory performance and naming ability were significantly poorer in the elderly. Although performance at the lowest level of intensity was somewhat poorer for the elderly, relative to their medium and strong levels, there was no significant main effect of strength or interaction with age, suggesting (1) that this method of compensation did not ameliorate the effects of ageing and (2) that threshold and intensity differences can not wholly account for reduced recognition and naming performance in older participants.

Although threshold (i.e., detectability) typically increases with age, this cannot account for all of the decline in identification performance, suggesting some degree of dissociation between them. A further question then arises as to the extent to which identification deficits are a consequence of (1) impoverished discriminative ability, (2) degraded semantic memory, and

(3) poorer lexical access. Larrson and Backman (1997) argue that degraded semantic memory is unlikely as the decline in visual object identification with age is far smaller than that observed in equivalent olfactory tasks, leaving discriminability and lexical access as the most likely alternatives (see Larrson and Backman 1997).

A role for impoverished discriminative ability in normal ageing is suggested by two other findings in healthy elderly people. First, Schemper, Voss, and Cain (1981), screened elderly (mean age = 76) and young (mean age = 18) female participants on their ability to discriminate odors, by presenting them with a name (e.g., garlic) followed by presentation of three different odors, one of which was the target. Participants had to identify the target odor. Using 20 such trials, they reported that 45% of the elderly failed this test, compared with just 4% of the young participants. Second, Schiffman (1992), as mentioned above, collected threshold and naming data from 143 participants of varying ages (10–80). However, in addition, Schiffman (1992) also measured discriminative ability (the triangle or oddity test) and odor similarity. Because there were no threshold differences between age groups, any effects are unlikely to be attributable to this cause. Schiffman (1992) found that mean percent correct on the discrimination task fell somewhat in the sixth decade (70% correct) and dramatically in the seventh decade (45% correct). Similarity ratings were analyzed using multidimensional scaling and this too revealed that ageing altered the relative position of odors in odor-space, mainly in the sixth and seventh decades, indicating that these participants had difficulty in judging the degree of qualitative similarity between different odors (see fig. 6.5). These results suggest, along with the others reviewed earlier, that with advancing age the memory system responsible for odor object recognition declines, resulting in poorer ability to tell different odors apart.

As in research on normal ageing, the study of olfactory ability in Alzheimer disease also suggests a distinction between detectability and other functions (see Martzke, Kopala, and Good 1997). Although several published studies have revealed heightened olfactory thresholds (e.g. Doty, Reyes, and Gregor 1987; Morgan, Nordin, and Murphy 1995), an equally significant number of have failed to do so (e.g., Rezek 1987; Koss et al. 1988; Morgan and Murphy 2002). At least some of these differences in threshold findings between studies probably result from patient heterogeneity, in that those with more severe symptoms are more likely to have severe detection problems (Murphy 2002). However, *every study* has observed significant deficits in higher olfactory functions, especially in the ability of patients with Alzheimer

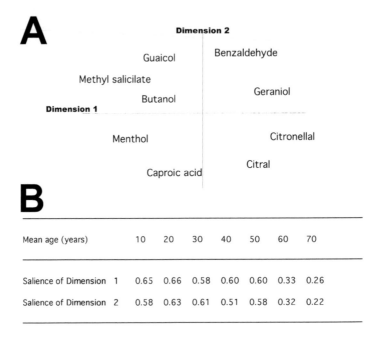

A

Dimension 2

Guaicol · Benzaldehyde

Methyl salicilate

Butanol · Geraniol

Dimension 1

Menthol · Citronellal

Citral

Caproic acid

B

Mean age (years)	10	20	30	40	50	60	70
Salience of Dimension 1	0.65	0.66	0.58	0.60	0.60	0.33	0.26
Salience of Dimension 2	0.58	0.63	0.61	0.51	0.58	0.32	0.22

Fig. 6.5. (A) Multidimensional scaling solution (two dimensions) for nine odors smelled and judged for similarity by participants aged from 10 to 70 years. (B) Salience of the two-dimensional solution above by decade, indicating that older participants, those in their sixties and seventies do not perceive these odors in the same way as younger participants (data adapted from Schiffman 1992).

disease to name odors (see Mesholam et al. [1998], for a meta-analytic review). A disassociation between detectability and naming is further suggested by the failure to observe a correlation between detectability and naming performance in the only two studies that reported looking for it (Koss et al. 1988; Morgan, Nordin, and Murphy 1995).

To what extent then do the observed deficits in odor naming reflect loss of discriminability? As before, deficits might also be caused here by degraded semantic memory or impaired lexical access. Larsson et al. (1999) found that providing verbal and visual prompts *did not* improve performance on odor identification to a level equivalent to elderly controls, suggesting that deficits in lexical access alone were unlikely to be producing poorer odor naming in their early-phase Alzheimer group (see Rezek [1987] and Morgan, Nordin, and Murphy [1995] for conceptually similar findings). This difference could not be attributed to poorer detectability either, as the groups did not significantly differ in this respect. The other alternative of degraded semantic memory was also excluded. Larsson et al. (1999) found that odor identification was not correlated with the Boston Naming Test, although the relationship was positive ($r = 0.3$). If degraded semantic memory solely accounted for poorer odor identification, then a strong relationship between these two tasks would be expected and this was not observed. This leaves impoverished discrimination as the most likely alternative.

To date, there has been no direct investigation of odor discrimination in the Alzheimer literature. Although delayed matching to sample with odors (Kesslak et al. 1988) and odor recognition memory is impaired (Murphy, Nordin, and Jinich 1999; Nordin and Murphy 1996) the extent to which these processes are affected by discrimination loss alone can not be determined. However, Moberg et al. (1987) reported that *prior* to testing recognition memory in an Alzheimer group, all participants were screened for discriminative ability. Unfortunately, no data is provided about the nature of the screening test nor about the number of participants that were subsequently excluded. Their findings suggest that deficits in recognition memory in Alzheimer disease, which they observed, can not be solely attributed to discrimination loss.

Several different conditions, notably temporal lobe epilepsy, brain injury resulting from cardiovascular accident, and frontal and temporal lobectomy, can all result in deficits in odor identification and discrimination, independent of changes in detectability. These studies can be broadly grouped into three categories: first, those dealing with identification; second, those dealing with discrimination, but in which adequate discrimination controls have not been instigated; and third, those dealing with discrimination, but in which attempts have been made to exclude task specific problems. All these findings point to a common conclusion, that deficits in quality discrimination can occur and that these effects may be obtained independently of deficits in odor detection.

Excisions of the temporal lobe and the frontal lobe impair odor identification as do cerebrovascular accidents to the temporal lobes (Jone-Gotman and Zatorre 1988; Savage et al. 2002). In each case olfactory detection thresholds were normal and brain-injured controls did not evidence olfactory deficits. Although the literature clearly indicates that discriminative ability plays an important role in naming odors, this was especially evident in a study of 17 patients with temporal lobe epilepsy conducted prior to surgery (Eskenazi et al. 1983). Discrimination was assessed using a triangle test and, overall, discriminative ability was significantly poorer in these patients (mean = 71% correct) than in normal controls (mean = 87% correct). Although detectability (threshold) did not significantly correlate with discriminative ability, the epilepsy patients showed a substantial correlation between discrimination and naming ($r = 0.8$) and discrimination and recognition memory ($r = 0.5$). Both recognition memory and naming were also significantly impaired. Similar discrimination deficits, independent of detection, have been observed following temporal lobectomy (Martinez et al. 1993), after temporal and frontal lobectomy (Zatorre and Jones-Gotman, 1991), and following frontal lobe damage (Pol et al. 2002). Although these all clearly show discrimination problems independent of detection, these studies cannot rule out the possibility that deficits in discrimination arose because of some task-specific difficulty.

Such an account is unlikely in the final few studies reported here, as these studies used various controls to discount this possibility. Potter and Butters (1980) examined whether discrimination deficits using a same-different procedure would be evident for participants with damage to the orbitofrontal cortex, with Korsakoff syndrome and in brain-damaged and normal controls. In addition to measuring olfactory threshold, the study also assessed hue discrimination by using a task analogous to that used with the odors. Although patients with orbitofrontal lesions and Korsakoff syndrome were highly impaired on discrimination ($d' < 0.5$), normal and brain-damaged controls performed equivalently. There were no deficits in threshold in the orbitofrontal or brain-injured groups relative to normal controls, but Korsakoff patients did have elevated thresholds. More importantly, hue discrimination was largely normal, except being somewhat depressed in Korsakoff patients. This failure to observe a difference was not an artifact of test difficulty, as both the olfactory and hue tests were graded and this did not systematically affect performance. This study clearly shows that participants can be found who are able to normally detect odors, but who are extremely poor at qualitative discrimination, yet who are able to perform the task of discrimination in another modality without evident impairment.

Korsakoff patients, as suggested earlier, show deficits in odor discrimination and on related tasks in other modalities too (see Mair et al. 1986). The most comprehensive study of their ability in respect to olfaction, examined olfactory discriminative ability using a delayed matching to sample (DMTS) test, in which both odor similarity and delay (5, 15 or 30 seconds) were manipulated (Mair et al. 1980). There were two control conditions. The visual condition involved seeing a face, then attempting to pick the face from a set of faces, simultaneously, successively, or after a 15-second delay. The second control procedure was a consonant trigram task, in which participants were asked to recall the consonants after delays of 0, 3, 9, or 18 seconds. Odor threshold was also assessed and there was no significant difference relative to controls. The first finding here was that odor discriminative ability, as measured by DMTS, was significantly worse in the Korsakoff group than in controls, especially where the target and distractor were highly similar. The crucial finding, however, was that performance did not change over the varying delays (see fig. 6.6), implying that the deficit did not result from an inability to retain olfactory information across the DMTS interval. The second finding was that facial identification was similar for Korsakoff and normal participants when presented simultaneously, but that performance in the Korsakoff group dropped markedly as the delay increased, whereas the performance of healthy participants did not (see fig. 6.6). The third finding, similar to the faces task, was that recall of the consonants was similar to controls upon immediate recall but decayed significantly faster for patients with Korsakoff syn-

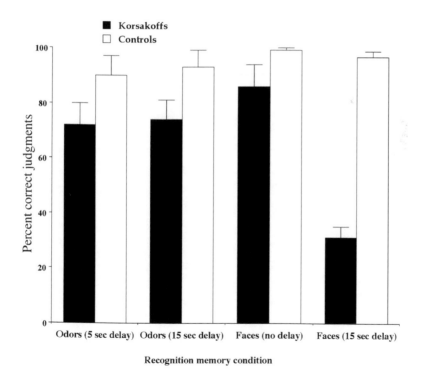

Fig. 6.6. Mean percent correct recognition memory score (standard error) for faces and dissimilar odors, from patients with Korsakoff syndrome and controls (data adapted from Mair et al. 1980).

drome. Apparently, olfactory discrimination deficits in Korsakoff patients involve a failure to discriminate olfactory stimuli that is independent of both detectability and retention interval. Moreover, the ability to discriminate faces and recollect trigrams almost as effectively as healthy participants at zero delay, suggests that inability to perform this type of task is not the source of the olfactory deficit.

The final and perhaps most compelling example is the single case study concerning HM, who received a bilateral temporal lobectomy for intractable epilepsy (Eichenbaum et al. 1983). HM showed normal olfactory thresholds for a variety of stimuli. Using a paired-comparison procedure, HM was able to discriminate between the same odor presented at higher and lower concentrations as well as control participants. However, when the odors only differed by quality, HM performed at chance level, on both easy, intermediate, and hard qualitative odor discriminations. This result is important for two reasons, first, because HM was unable selectively to perceive odor quality, in so much as his ability to discriminate odors differing primarily on this variable was lost. Second, he was clearly able to perform the discrimination task when the odors differed only in intensity, suggesting that a task-specific deficit could not account for these findings, nor could inadequate ability to detect odors.

Although the neuropsychology data offer some striking support for dissociations between detectability and qualitative discriminability, there have been almost no investigations into abnormalities of odor hedonics. Only two types of finding are currently available in the neuropsychology literature. The first concerns Alzheimer disease. Perl et al. (1992) found that patients with Alzheimer disease were able to discriminate pleasant and unpleasant odors (based on judges coding their facial expressions). Although no data on identification were obtained, Perl et al. (1992) concluded that appropriate hedonic reactions could occur to pleasant and unpleasant odors in the absence of identification—a not unreasonable assumption because these patients had largely lost the ability to verbally communicate. Soussignan et al. (1995) found intact hedonic responding to pleasant and unpleasant odors, also judged by facial expression, in low-functioning mute children with pervasive developmental disorder. In both these cases it seems reasonable to presume that verbal odor identification could not have played a significant part in the hedonic responding, suggesting an independent hedonic system.

A second and more recent discovery is the selective deficit in the recognition of facial displays of disgust in Huntington disease (e.g., Sprengelmeyer et al. 1997). Not only are the identification of disgust faces affected, but so too are disgust responses to odors found pleasant or unpleasant by healthy participants (based on ranking odorous stimuli [Lawson, Stevenson, and Coltheart, in preparation]). However, many patients with Huntington disease show deficits in olfactory identification, which appear to result from a progressive inability to detect odors (Nordin, Paulsen, and Murphy 1995). Consequently, the degree to which abnormal hedonic responses in this disease result from selective impairment in affective responding versus loss of olfactory sensitivity is not currently known. The evidence, overall, little as there is, suggests a possible dissociation between hedonic responding and other olfactory functions.

Neuroimaging

The neuroimaging data has made three important contributions to understanding human olfaction. First, it has allowed the mapping of function to neuroanatomy in intact individuals. Second, it has resulted in a better understanding of the way in which olfaction relates to other sensory systems. Third, it has resulted in some preliminary statements about dissociations between different olfactory functions and the information processing hierarchy. The first and second sets of findings are not directly relevant to this section because they pertain directly to neuroanatomy and are covered in the earlier chapters of this book. The third set of findings is relevant; they are discussed below.

Neuroimaging studies of differing olfactory functions have revealed four major findings. First, there is a dissociation between brain areas involved in olfactory hedonics and those involved in other measured functions (e.g., familiarity, edibility, detection; Royet et al. 2001). Second, emotive odors activate brain regions distinct from those activated by emotive pictures and sounds (Royet et al. 2000). Third, and relatedly, there is a dissociation between brain areas activated by odors reported as pleasant and those reported as being unpleasant (Rolls, Kringelbach, and de Araujo 2003; Gottfried, O'Doherty, and Dolan 2002). Fourth, as tasks become more complex, the range of structures involved in processing broadens from those known from neuropsychological and animal studies to be primarily olfactory, to structures that are believed to engage in multimodal functioning (e.g., Royet et al. 2001; Qureshy et al. 2000). What is important here, in particular, is the finding of brain areas for which information processing is concentrated on the emotive aspects of olfaction, rather than solely on the perceptual, as this supports the general supposition here that hedonic and perceptual (quality/intensity) systems represent separate information-processing streams in the olfactory system.

Top-down Processing

Except under the rather unusual conditions that pertain in the odor laboratory, most smells occur in tandem with other sensory cues. The smell of orange typically co-occurs with the sight of an orange, toilets that look filthy probably smell bad too, and the sight and sound of a crackling fire likely smells smoky. As discussed below, such cross-modal cues have a demonstrable effect on participants' reports of olfactory experience, suggesting that top-down processing can affect what is reported under certain conditions. However, although the evidence reviewed below plainly suggests that top-down modulation occurs, the key question remains as to whether the result is a change in the olfactory percept or in the way that the olfactory percept is reported.

Under various conditions, color has been shown to affect odor identification, intensity, and hedonic judgments. Engen (1972) used a signal detection paradigm to examine whether colored blank stimuli would affect false-alarm rates when participants were asked to judge whether an odor was present or absent. Colored stimuli, that had no odor, were more likely to be judged as odorous than uncolored blanks. Color can also affect the qualities that are reported when participants consume colored and odorous stimuli. Using both odor typical (yellow color/lemon odor) and atypical colors (orange color/ cherry odor), Dubose, Cardello, and Maller (1980) found that both conditions increased the perceived intensity of the odor, whereas atypical colors (e.g., orange-colored cherry-flavored drinks) induced odor ratings that were characteristic of the odor normally associated with that color. Identification of an odor is also affected by color. Zellner, Bartoli, and Eckard (1991), found that typical colors enhanced identification, but that atypical colors had no effect. The most dramatic demonstration of the effect of color on odor judgments are those of Morrot, Brochet, and Dubourdieu (2001). They found that qualitative evaluations of a white wine colored red by an odorless and tasteless colorant were more similar to those of a red wine, than they were to either the *same* white wine, or to the *same* colored white wine presented blind.

At least for judgments of intensity, color-based effects do not appear to arise from the demand characteristics of the situation. Zellner and Kautz (1990) presented participants with a standard odor, which for half the participants was colored and for the other half not. In each case, participants were asked to judge whether a comparison stimulus was stronger or weaker — that is no direct ratings of intensity were obtained. For participants receiving a colored standard, all the comparison stimuli were uncolored, whilst for those receiving the uncolored standard, all the comparisons were colored. Consistent with other findings (e.g., Zellner and Kautz, experiments 1 and 3), the colored standard tended to be judged as stronger than an equally concentrated comparison stimulus, whereas for participants with the uncolored standard, the equally concentrated but colored comparison, was more often judged the stronger.

Verbal labels can also affect odor judgments. Herz and Clef (2001) selected a set of odors based on their potential hedonic ambiguity. Labels exerted a significant impact on whether the odors were judged favorably or unfavorably. For example, a mixture of isovaleric and butyric acid described as Parmesan cheese in one condition was liked more, than when it was described in another as vomit. Clearly demand might play a significant role in this sort of experiment, but their observation of a first-label effect makes this appear less likely. As the design was within-participant, both labels for an odor were presented at different times and in a counterbalanced order across participants, thus a first-label effect was detected based on between-participant comparisons. Herz and Clef (2001) observed that in certain cases (for pinene and menthol), the first label exerted a greater effect than a second hedonically different label. That there was carry-over suggests that participants were doing something more than simply responding appropriately to each label.

Labels and colors then, clearly influence how participants report an odor's smell. A key question is whether such effects influence perception or whether they result in responses consistent with the more dominant information channel (typically vision/audition). That is does orange-colored lemon odor really smell *like* orange or is it just *reported* as smelling like orange? Although this question is a hard one to address, the evidence reviewed above is consistent with both accounts and offers no clear guide as to whether top-down processing actually shifts odor perception or the propensity to report information consistent with the dominant sensory channel. There is, however, one experiment that does differentiate between these accounts. Gottfried and Dolan (2003) explored whether congruent olfactory-visual information (i.e., diesel odor and a picture of a bus) would enhance response latencies to the question "is an odor present?," relative to both incongruent olfactory-visual information (i.e., fish odor and a picture of cheese) and to a condition in which odors were presented alone. The shortest latencies were obtained when the visual and olfactory information was congruent. There was no significant difference in latencies between the incongruent and odor-alone conditions. Gottfried and Dolan (2003) concluded that the failure to obtain a difference between the latter conditions reflects a decision based on semantic associations between the odor and image, as misperception might have been expected to *boost* latencies under both visual conditions, but not when an odor was presented alone.

Motivation

Motivational variables appear to affect participants' hedonic reactions to odors, but little research has been conducted on their impact on the qualitative and intensive dimensions. Two types of finding have been suggested in the literature, but at the moment only the first has firm empirical support: (1) hunger and satiety can alter hedonic responses to food odors but not to nonfood odors, and (2) sexual arousal may increase the acceptability of certain body odors. Several studies have now shown that after a meal, food-related odors are rated as less pleasant, an effect that gradually ameliorates over time. This effect can be mediated by the presence of food in the gut, such that following ingestion of a meal or a sucrose preload, food odors are rated as smelling less pleasant than they did before ingestion (within-participant) and less pleasant than a fasting control group (Cabanac 1971; Duclaux, Feisthauer, and Cabanac 1973). In addition, a further effect, sensory-specific satiety can be produced solely by repeated exposure to a food odor, but without ingestion (Rolls and Rolls 1997) and this effect is independent of any change in the perceived intensity of the target odor. In both these cases it is assumed that the qualitative aspects of the olfactory sensation remain unchanged, whereas only the hedonic component diminishes, but this has yet to be shown definitively (but see preliminary findings reported earlier in this chapter).

Sexual odors, that is, those volatiles associated with the organs of procreation, are typically regarded as unpleasant (Daly and White 1930). It has been argued, however, that this reaction also depends on motivational state (Kalogerakis 1963) and that sexual arousal leads to a change in the way in which these odors are hedonically evaluated. Although this makes good sense, as does a similar notion that motivational state may render previously disgusting food palatable (e.g., a rotting bread crust to a starving man), no empirical evidence as yet exists. Some weak anatomical confirmation for this idea has been obtained from neuroimaging data, in that romantic love primarily activates the insula (Bartels and Zeki 2000), a structure with known importance in mediating the emotion of disgust (Phillips, Young, and Senior 1997; Davidson and Irwin 1999).

Conclusion

The most striking characteristic of human odor perception is its plasticity. This plasticity is expressed in many ways: qualitatively, in that learning and experience can modify the way odors are reported to smell, their similarity to each other, and their discriminability and hedonics, in that motivational state and past experience can alter whether an odor is liked or disliked. This plasticity is entirely inconsistent with the view that odor perception is based primarily on the structural description of the odor, that is, its physicochemical properties. The plasticity of odor perception is, however, entirely consistent with the view that odor perception is a form of object recognition. First, such a process depends on learning, in that patterns of features that occur together have to be bound and retained so that the pattern can be recognized and its consequences evaluated. Second, the absolute reliance of object recognition is demonstrated by the neuropsychological data, which show a constellation of deficits characterized by an ability to detect odors, but an inability to discriminate between them.

After the object recognition process has taken place, the hedonic reaction to the odor occurs, although there is a limited body of neuropsychological evidence that suggests that affective reactions can occur independently of this process (possibly in Alzheimer disease and pervasive developmental disorder). Notwithstanding this caveat, as indicated above, this also involves striking plasticity, and the hedonic response to an odorant can be seen as its most important consequence if it is reasonably assumed to be the behavioral correlate of approach/avoidance. A further finding is that odor perception is heavily affected by ongoing information from other sensory channels—top-down processing. However, what is so intriguing here is that we do not currently know whether olfactory perception per se is brought into line with the other senses (when they mismatch) or whether all that changes is the report that participants provide about the stimulus.

In sum, functionally, human odor perception requires that complex stimuli (objects) be recognized, not features, and that participants are enabled to discriminate these objects and react appropriately to them (approach/avoidance). The empirical evidence suggests that these functional goals are met by a process of object recognition that is heavily dependent on learning and memory, and by an hedonic reaction that also demonstrates considerable plasticity, both through short-term means (motivational state) and longer-term means (evaluation of consequences). Notwithstanding this, the system demonstrates a further flexibility, in that when it is out of synchrony with information from other presumably more dominant sensory channels, it is the olfactory system which is realigned. Whether this realignment is perceptual or semantic is not currently known.

CONSTRAINTS IMPOSED BY THIS SOLUTION

If the primary role of the human olfactory system is to construct and retain odor objects out of complex mixtures of chemicals, so that these objects can be more readily discriminated from the chemosensory background, this implies certain limitations to the way in which humans can perceive odors. The most obvious limitation is that it should affect our ability to identify components of complex odor mixtures, as the system output is in wholes not parts. If such a limit is imposed by the perceptual system, it should be apparent in all individuals, including perceptual experts. Three further predictions can also be made. First, if olfaction depends on pattern recognition, then the stored patterns must be relatively resistant to interference and trace decay. Not only would this prevent the loss of valuable prior experience but, more importantly, it would also enhance discriminability by preserving psychological distance between odors. That is, encoding everything would lead to so much inherent similarity between odors that discrimination would become very hard as a consequence of overgeneralization. Second, once an odor object has been acquired, even rudimentary components of it may be enough to recognize it. This would allow recognition of even the remnant scent of a predator and, as described earlier, the characteristics of tastes originally experienced with that odor (i.e., sweet-smelling odors). Third, concerning adaptation, to assist in identifying odor objects the olfactory system must adapt relatively fast to the olfactory background, thus allowing any new stimulus to come to the fore. Consequently, the olfactory system should demonstrate rapid adaptation. All of these points, with the exception of adaptation, which is discussed elsewhere, are examined next.

Limitations on Identifying Parts

Three lines of evidence converge on the conclusion that humans are limited in their capacity to identify the component parts of odor mixtures. Laing and colleagues performed the most complete set of studies. Their basic technique involved familiarizing the participant with a set of seven labeled odors and then presenting the participant with either a single odorant or mixtures composed of two or more components drawn from this set. The participants' task was then to identify the odorant/s present on that trial. Laing and Francis (1989) found that participants were able to identify, correctly, single odorants on 82% of trials, the components of binary mixtures on 35% of trials, three-component mixtures on 14% of trials, four-component mixtures on 4% of trials, but they were never able to accurately identify all the elements of five-component mixtures.

Obviously, this task is a hard one, especially for the more complex mixtures as it involves both remembering the odor names, matching the names to the odor, and identifying each individual odor in the mixture. Taking this into account Laing and Glenmarec (1992) conducted a further experiment in which participants were familiarized with the target odors and then, prior to every trial, they were asked to smell a target odor and they then had to judge whether the target odor was present or absent in the stimulus (again composed of single odors and multicomponent mixtures as described above). In addition, the same participants were also asked to identify all the components using the method described in the preceding paragraph. This procedure produced virtually identical results in both conditions, that is, performance dropped markedly as the mixtures increased in complexity (see fig. 6.7).

It is well known in the flavor industry that some odors blend together well and that others do not. Therefore performance on these types of tasks might be severely limited by the choice of odors in that good blenders might reduce identification of components and poor blenders increase it. Using good- or poor-blending odors selected by industry experts, Livermore and Laing (1998a) assigned participants to one of these two sets of stimuli. The results indicated that poor blenders were better identified when mixtures contained less than three components. However, both good- and poor-blender groups were *equally* impoverished with higher order mixtures.

If odors are treated as objects, then it might be expected that participants would be as good at identifying a target *object* in a mixture as they are at iden-

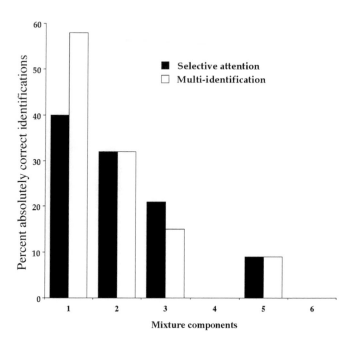

Fig. 6.7. Mean percent of absolutely correct selections on the selective attention and the multi-identification tasks (data adapted from Laing and Glenmare 1992).

tifying pure chemical stimuli and their mixtures. Livermore and Laing (1998b) investigated this possibility using several odor objects, each composed of a large number of chemicals (e.g., chocolate, rose, and honey). The same technique employed in their earlier studies was used, in which participants were presented with mixtures of these odor objects, their task being to identify which odors were present. These findings are of especial interest (see fig. 6.8), because they show that the *same limitations* apply to complex odor objects as they do to single pure chemicals, in that absolutely correct identification performance drops to chance level when there are more than four or five components.

If the ability to identify odor components in mixtures is a fundamental aspect of olfactory processing, then it should not be possible to exceed this limit even if you have received training to do so. Nonetheless, below this perceptual barrier it should be possible to enhance performance through training. Livermore and Laing (1996) investigated this by comparing an expert group of perfumers and flavorists against a group of untrained participants. Each group had the same task, namely to identify the elements present in multi-

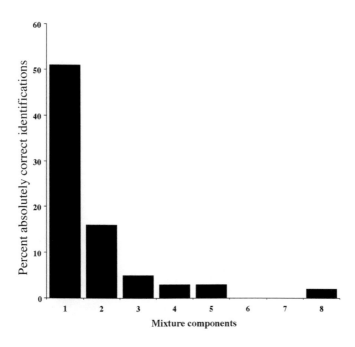

Fig. 6.8. Mean percent of absolutely correct selections for multicomponent odor objects (data adapted from Livermore and Laing 1998).

component mixtures. Experts were significantly better at identifying the components of two- and three-component mixtures than the novice group. However, with four- or five-component mixtures performance was uniformly poor (see fig. 6.9), suggesting that this limit reflects a genuine perceptual barrier, rather than a lack of expertise in the participants utilized in their other experiments.

Another approach to studying whether participants can detect components in odor mixtures is to ask them to judge the complexity of the mixture, either through simply stating the number of odors present or through rating complexity on a rating scale. Presumably, if participants are aware of the increasing number of components, complexity judgments should increase linearly with mixture complexity. On the other hand, based on the findings reviewed above, it would be expected that complexity judgments would increase for mixtures up to three or four components, but with no further increase beyond this point. Three studies (Laing and Francis 1989; Laing and Glenmarec 1992; Moskowitz and Barbe 1977) adopted the first measure of complexity, using number of odors judged to be present. As expected, this in-

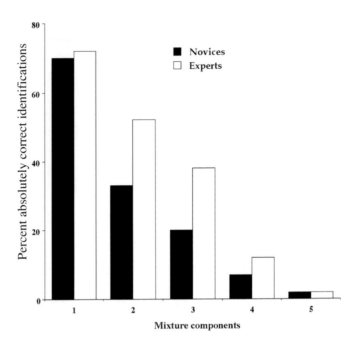

Fig. 6.9. Mean percent absolutely correct selections for normal participants (novices) and olfactory experts (data adapted from Livermore and Laing 1996).

creased up to a ceiling of three components, suggesting that participants not only were poor at discerning components irrespective of identification accuracy, but also indicating that little feature-based information could be present (contrast this with a visual example of progressively adding pen strokes to a small piece of paper, complexity emerges here directly through observing the increasing number of pen strokes/features). Only one study has examined direct ratings of odor complexity. It is arguable whether this study assesses the same thing as reliance upon number of components, because participants may judge complexity on the basis of redolence to other odors (i.e., what it smells like). Jellinek and Koster (1979) found no relationship between the number of components present in a mixture and participants' complexity ratings. If an odor's perceived quality does depend on matching to a memory-based system, even a single pure chemical may be redolent of many different qualities — that is, judged complexity should be independent of the stimulus if it is based on experience.

Finally, if participants are unable to identify the components of more complex mixtures, this should also affect the similarity of the mixture to the constituent components. Jinks and Laing (2002) explored this by having participants evaluate each single odorant or mixture using the 146-item Dravnieks scale. As the number of elements present in a mixture increased, there was little change in the number of reported odor qualities, but the type of odor quality reported shifted from the specific to the generic. As unfamiliar odors are both hard to name and to discriminate, this shift to more generic descriptors suggests that participants tended to treat more complex mixtures as novel stimuli rather than as a set of familiar components.

Constraints on Forgetting

To be able to distinguish between odors, previous experience with those odors needs to be retained. If every olfactory experience were encoded, this would produce catastrophic interference, resulting in one odor reminding a participant of every other experience; consequently, some brake is needed on this process and this may be manifest as strong resistance to interference relative to other forms of learning. Evidence of high resistance to interference has been obtained both with odor-taste learning and with odor-odor learning as detailed below.

Resistance to interference in odor-taste learning has been explored using two different interference procedures. The first extinction involved presenting participants with three odors, two of which were paired with citric acid and one with sucrose (Stevenson, Boakes, and Wilson 2000a). In a second phase, participants received one of the citric acid–paired odors alone in water (i.e., the extinction procedure). This was followed by a posttest in which all three odors were rated on a series of scales, including how sweet and sour they smelled. No evidence of extinction was obtained, in that the citric acid–paired odor, subsequently presented alone in water, was judged to smell as sour as the citric acid–paired odor that had not been extinguished and with both smelling more sour and less sweet than the sucrose-paired odor (see fig. 6.10).

The conclusion that no interference had occurred did not rely on a null result. Participants were also asked within the same experiment to learn the relationship between colors and tastes (e.g., red sucrose). Two colors were paired with citric acid and one with sucrose. One of the citric acid–paired colors was then presented alone in water (i.e., the extinction phase). Using a test of conditioning which could assess both odor-taste and color-taste learning (look and smell the solution and predict what it would taste like) indi-

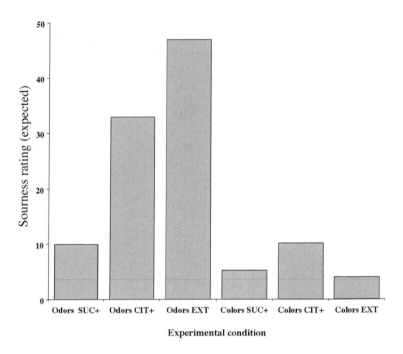

Fig. 6.10. Median expected sourness ratings for odor-taste pairings (sucrose paired [SUC+], citric acid paired [CIT+], and citric acid paired then water paired [EXT]) and color-taste pairings (same legend) (data adapted from Stevenson et al. 2000a).

cated that although participants now expected the extinguished citric acid color to *not taste sour*, the procedure had not exerted any effect on the extinguished citric acid–paired odor. Thus, within the same procedure, a color-taste association had been affected by the procedure, but an odor-taste association had not.

The second approach used an alternate interference procedure — counterconditioning. In this case three odors were again paired with tastes (two with citric acid and one with sucrose) after which one of the citric acid–paired odors was then paired with sucrose — hence, counterconditioning (Stevenson, Boakes, and Wilson 2000b). An analogous color-taste control condition was again employed. Use of the expectancy test described above to simultaneously assess learning in the color and odor conditions revealed that, although color-taste learning was affected by the counterconditioning, odor-taste learning was not.

Presentation of the odor alone following conditioning apparently exerts little effect on the acquired taste characteristics of the odor. The converse of this procedure also appears to be true, in that presentation of the odor prior to pairing with a taste prevents such associations forming (Stevenson and Case 2003)—a corollary of the retroactive interference procedures described above (i.e., proactive interference). In this case, participants were exposed to two odors, four times, prior to a conditioning stage in which one of these odors was then paired with sucrose and the other with citric acid. In addition, participants also experienced a further two odors, one of which was paired with citric acid and the other with sucrose. The odors presented only in the conditioning phase acquired the appropriate characteristics, that is, the sucrose-paired odor got to smell sweeter and the citric acid–paired odor sourer. No evidence of conditioning was obtained in the odors that had been exposed prior to pairing them with sweet or sour tastes. This suggests that it is the exposure process per se that results in encoding and that once this has happened little further learning takes place (or at least none could be detected).

This characteristic of resistance to interference in odor-taste learning is also shared by odor-odor learning. In a recent series of experiments we have begun to explore this using two different types of procedure. The first, analogous to the extinction procedure described for odor-taste learning, involved presenting participants with two-odor mixtures (AX, BY), followed by presentations of the elements of one of the odor mixtures alone (B, Y), several times (Stevenson, Case, and Boakes 2003). This procedure did not affect performance on a subsequent similarity test, in which participants were asked to judge the similarity of the two odors mixed together (A vs. X), versus the two odors mixed together and subsequently smelled alone (B vs. Y) versus a new set of two odors not presented in the experiment (C vs. Z). Here, the odors that had been originally paired together did not differ in acquired similarity, but both reliably differed from the nonpaired, nonexposed controls (C vs. Z). Thus, as before, no evidence of interference could be detected.

As with the odor-taste data, a significant concern is not to rely upon a null result. This issue was dealt with by having participants learn associations between odors and colors (e.g., red-mint) in the same experiment. One of these combinations was then presented with each element alone (e.g., red water, mint water). To test learning a new procedure was devised, capable of measuring simultaneously the odor-odor and odor-color learning. This was based on the participant making frequency of occurrence estimates for all the stimuli alone (colors and smells) and for all possible mixtures. These ratings readily revealed that participants were aware that the extinguished color and odor had occurred more frequently alone, than as a mixture—that is, participants had correctly tracked the experimental contingencies. This was not evident for the extinguished odor-odor mixture elements. This is because if learning were preserved, each element would be confused with the other and with their mixture, because they should all smell somewhat similar as a result of the earlier learning episode. As a result, similar frequency estimates for all three stimuli (e.g., A, Y, and AY) should be observed and this was in fact the case for *both* the extinguished and the nonextinguished but paired conditions. Thus participants on this test demonstrated an awareness of color-odor extinction but not of odor-odor extinction.

Repeated presentation of single odors alone, as described above, especially prior to the test phase, must raise a concern about adaptation. This might lead to the extinguished odors being judged as more alike, simply because participants could not smell them adequately. To explore this we set out to examine what would happen to two odors if they were *just* repeatedly presented during a mock interference phase, relative to (1) two odors that had been experienced as a mixture (but with no interference) and (2) two odors presented for the first time on the test phase (Stevenson, Case, and Boakes 2003; experiment 2). Similarity ratings taken during the test phase revealed two interesting findings. First, presentation of the two odors during the mock interference phase did not render the pair as more alike, relative to the nonexposed controls or the paired controls. That is, adaptation could not account for the absence of an interference effect in the experiment reported above. Second, exposure during the interference phase actually resulted in the two odors being judged as significantly more dissimilar than the nonexposed controls, consistent with other findings, which, reported earlier, indicate that exposure alone can enhance discriminability between odors.

More recent studies have explored a further interference phenomenon, in which, following the pairing of two odors, say A and X, A is then subsequently paired with Y, and X with B. This classic interference design also demonstrates resistance to interference of a magnitude similar to that observed with the extinction-like procedure described above (Stevenson and Case, submitted). In a further experiment we tested whether any learning had taken place for the interfering odor pairs, namely AY and BX in the example above. It had, when contrasted against a nonpaired control odor on the similarity test, but learning was significantly less evident, than in the original paired condition (AX above), suggesting, as observed in other studies (see below), proactive interference.

Several other lines of inquiry have suggested that odor memory is resistant to interference. Typically these have involved one of two approaches. The first is paired associate learning between odors and pictures, which demonstrates proactive interference but not retroactive interference (Lawless and Engen 1977). The second approach has examined whether participants' ability to recognize which odors have been previously presented when targets and distractors are presented at varying times after training (e.g., Lawless and Cain 1975). The typical finding is of a relatively flat, forgetting curve over time, relative to recognition memory for pictures that decays relatively fast. Memory for hard-to-name shapes was, however, substantially similar to that of the odors (Lawless 1978). Both these types of study, although frequently cited as examples of odor memory strength, are probably not *directly* related to the effects reported above of the persistence of odor-taste and odor-odor associations. First, there are important methodological differences. The paired associate learning experiment of Lawless and Engen (1977) employed successive stimulus presentation, whereas all of our studies employ simultaneous presentation. In addition, Lawless and Engen (1977) make no claim as to the discriminability of the stimuli following learning or to changes in odor quality, because their study and the others cited above rely on recognition as the dependent variable. As we suggest later in this book, recognition memory may rely on the perceptual memory system that we believe underpins odor-taste and odor-odor learning, but it does not automatically follow that this reliance produces effects synonymous with high resistance to interference in recognition memory tasks as there are clearly studies that do not find this. For example, Walk and Johns (1984), using a delayed matching to sample task, which presented an odor drawn from the same perceptual category as the two targets presented in the acquisition phase, significantly reduced performance when participants were asked to pick the target from four stimuli also drawn from the same set. That is, the distractor stimuli interfered with recognition memory.

Performance resulting from most human learning is context dependent. Responses evoked by a particular cue elicit behavior in one environment, but not in another. This reliance on context to control performance is well established in the animal learning literature, especially for extinction, where an extinguished cue can be reinstated (spontaneous recovery) when presented again in a different (novel) context. Thus, the same cue can elicit different behaviors, with performance controlled by the context. For the perception of odor quality, we might expect that context would be relatively unimportant, smoke should still smell like smoke where and whenever it is encountered. Although we might speculate that context should not dictate odor quality perception, there are at present no data on which to base this inference.

Just as the perceptual characteristics of an odor need to be protected from further encoding, so too do the hedonic aspects. We have already discussed how evaluative conditioning using odors and tastes evidence resistance to extinction but not to counterconditioning. This should come as no surprise as we might reasonably expect that the consequences of smelling an odor may change and that such changes should be encoded. Likewise, flexibility might also be expected to manifest in terms of contextual sensitivity. As we noted above, smoke needs to smell like smoke wherever it is encountered, but its hedonic consequences may have to differ depending on the context in which the odor is smelled (e.g., being awoken to the smell of smoke is probably unpleasant and very frightening; smelling a garden bonfire may be unpleasant but is not usually frightening). As with odor perception, there are no data as yet to evaluate whether context-specific learning does occur with liking. For evaluative conditioning, attempts to demonstrate this have met with little success (see De Houwer, Thomas, and Baeyens 2001), nonetheless, it would be highly surprising if context *did not* modulate hedonic responses to odors because how else do we account for the liking of odors such as Parmesan cheese when their major volatile constituents are essentially those of vomit?

Redintegration

There are several theoretical ways, none of which have yet been confirmed, that might account for the resistance to interference and suspected insensitivity to context of odor quality perception. One possibility is that, as a result of encoding odor objects, presentation of fragments (i.e., akin to the interference procedures used in many of our experiments) can act to redintegrate (or recover) the whole percept. Thus, what is perceived is not the fragment but the whole. Consequently, as learning typically requires surprise, that is some way of directing attention to the stimulus, the failure to notice any change from the presentation of the whole to the part results in no further encoding. One effect of this is to protect already encoded objects. Another is that it allows the system to identify objects that are broadly similar to the target item. In some cases, this results at least in a significant disparity between the properties of the stimulus and the resulting percept. In most cases this disparity is beneficial, because, as noted above, it confers resistance to interference and it allows information about the stimulus to be presented in a perceptual form; for example, that an odor smells sweet and is thus likely to be a source of calories.

General Conclusion

Perceiving an odor and reacting to it are heavily dependent on learning and memory. Novel odors are harder to discriminate than familiar odors and less liked too. Children are poorer at discrimination odors and react differently toward them than adults. Laboratory or culturally mediated exposure to odors can affect their quality and hedonics. Loss of such olfactory information, as in normal ageing, Alzheimer disease, Korsakoff syndrome, and temporal lobe lesions, results in intact detection but adversely affects discrimination and perception of odor quality. All these findings point to a system that relies on mnemonic processes—object recognition and learned hedonic reactions. Both may be independent systems, modified by top-down processing and motivational state. Both systems have constraints. The object recognition system appears to make component identification harder and produces marked resistance to extinction to prevent overgeneralization. Learned hedonic reactions are conservative but more flexible and probably context sensitive. In sum, human olfaction demonstrates a high level of plasticity—a finding that is very difficult to reconcile with a receptor-centric view of odor perception.

7

Odor Memory

To this point, we have been concerned with the role of memory in odor object recognition and hedonics. A whole range of other information-processing tasks exists that rely on memory, and the purpose of this chapter is to examine and interconnect them with the object recognition process that we have alluded to in the preceding chapters. We begin with an exploration of human olfactory cognition, including short- and long-term memory, paired associate learning, odor priming, and imagery. We then explore many of these same issues in the animal literature and explore underlying neural correlates.

OLFACTORY COGNITION IN HUMANS

An important question one might ask is whether there is any need to attempt an integration of olfactory cognition, odor memory, and object recognition, and thus we start this section by explaining why it is necessary. Higher cognitive processes clearly influence perception—top-down processing—as we briefly described in the preceding chapter. For example, being able to name an odor appears to affect all three aspects of olfactory perception; a nameable odor appears more intense (Distel and Hudson 2001), it evokes a more potent emotional response (Herz and Von Clef 2001), and it is better discriminated too (Rabin 1988). Such findings can be conceived in two basic ways, either as perceptual or nonperceptual phenomena. Both alternatives might operate together or alone, and each could be envisaged in several different forms. Perceptual accounts might involve (1) the direct activation of an odor object encoding; (2) the refining of a representation, where the generation of a name for example, acts reciprocally to repress odor objects that are similar to the target's encoding; and (3) the enhancement of the target's level of activation. Nonperceptual accounts might involve (1) the recruitment of a second mode of representation (verbal), thus improving performance on memory and discrimination tasks (c.f., dual-coding; Paivio 1991); and (2) the alteration of the verbal description of a perceptual experience to bring it into line with information from the other senses. Whichever may be correct—and they are not mutually exclusive—top-down processing undoubtedly occurs (see earlier chapters) and we therefore need to know how the object recognition process that we have described is affected by it.

A second reason for integrating the process of object recognition with findings from olfactory cognition is that many cognitive processes in the visual system, for example, directly utilize perceptual pathways. The most notable of these is the capacity for imagery (Kosslyn and Thompson 2003). In the visual system, imagery calls upon the integration of a range of processes, most notably retrieval of perceptual encodings from long-term visual memory and their instantiation in short-term memory (Baddeley and Andrade 2000; Kosslyn, Ganis, and Thompson 2003). Important conceptual similarities exist between imagery and top-down processing, even though these are often treated as separate topics in the literature (Cain and Algom 1997). If imagery is conceived as a perceptual process, which many findings from visual neuroscience suggest (Farah 2000), then the generation of an image when an odor's name is provided, for example (imagine lemon), may involve virtually identical processes to those involved in top-down processing, where a name may act to refine or alter an extant perceptual experience. In other words the ultimate result of top-down processing may in certain cases be a perceptlike image. Two aspects of this are particularly important. First, if imagery is not a perceptual process as some have argued (Pylyshn 2003), this could imply that top-down effects are nonperceptual too (note the direct parallel with the discussion in the preceding paragraph). Second, several prominent olfactory theorists, especially Engen (1987), have argued that the olfactory system is incapable of imagery, whilst others (e.g., Cain and Algom 1997) suggest that imagery plays a key role in many olfactory tasks. Not only then do we need to understand how imagery (and top-down processing) may relate to object recognition, we also need to determine whether it constrains the type of explanation that we can offer for imagery and top-down processing. In essence does an object recognition account of olfactory perception suggest that images are perceptual or nonperceptual phenomena?

A third, and perhaps the most compelling, reason is that to date there is no account in the literature that presents a modal model of olfactory information processing. Rather, thinking in this respect has been guided, logically so in fact, by information-processing models derived from other modalities (c.f., White 1998) or from the study of human memory in general (c.f., Larsson 1997). Moreover, because there has been relatively little theoretical development in psychology of the basic information-processing steps in olfactory perception, cognitive processes have not typically been related back to perceptual ones, resulting in a lack of integration of the extant research literature. If this were not enough, perhaps the most striking finding of research into olfactory cognition during the past forty years has been that of differences with other modalities rather than similarities (Herz and Engen 1996; Stevenson and Boakes 2003). All this suggests that an integrated olfaction-centered model is needed. Such a model is inevitably going to be an approximation of the truth; nonetheless, it should serve a useful purpose if it helps to specify important questions to ask about olfactory information processing.

The main thrust of this section is human olfactory memory and related cognitive processes. We start by examining short-term memory for odors to see whether this differs qualitatively from the properties of long-term memory. Such differences do seem to exist in other sensory systems, do they in olfaction? Then we turn to examine the characteristics of long-term memory. To what extent are long-term memory findings synonymous with findings from memory processes involved in object recognition, more succinctly, are we talking about the same memory system or different memory systems? Clearly, when we smell something, other information may be available or become available, such as an odor's name or personally relevant memories, for example (Chu and Downes 2000a). How are these types of associations to odors formed and how are they linked to their object level description? We then examine imagery—the experience of an olfactory sensation in the absence of appropriate stimulation. Many have argued that olfactory imagery does not or cannot occur (see, for example, Engen 1987, 1991; Crowder and Schab 1995; Herz 2000). We examine the evidence pertinent to this claim and delineate how this might relate to olfactory information processing in general. Finally, we propose a model of human olfactory cognition, which attempts to integrate the findings reviewed here with an object recognition account of odor perception.

Short-term Memory

Four lines of evidence have been used in the past to argue for qualitative differences between long- and short-term stores-capacity differences, differential coding, serial position effects, and neuropsychological dissociations (White 1998). Using this framework, we examine the evidence here for differences based on these types of characteristics. Capacity differences do *appear* to occur. Engen, Kuisma, and Eimas (1973) used a delayed matching to sample procedure, in which participants either smelled one or five diverse odorants. After various intervals (3, 6, 12, and 30 seconds), a further odor was smelled, either the same or different to the one, or one of the five, presented earlier. When five stimuli were presented, rather than one, hit rates were significantly lower at all intervals, suggesting that over the short intervals examined, performance was limited to some extent by capacity, the difference being about 10%. A second study by Jones, Roberts, and Holman (1978) reached a similar conclusion. Here, participants smelled either one, three, or five odorants all drawn from a set of similar odorants (herbs). Participants then smelled eleven odorants, from which the one, three, or five had been drawn, and had to judge the degree to which they thought each of the eleven stimuli had appeared before. Hit rates fell as set size increased, from 72% for one odorant, to 62% for three odorants, to 58% for five.

To what extent do these findings reflect the operation of a qualitatively distinct short-term memory system? Two lines of argument suggest they do not. First, three studies have explored olfactory recognition memory over longer periods. Engen and Ross (1973) found that recognition for a set of 21 diverse odorants was about 67% (hit rate) at 30 days postexposure (see fig. 7.1). Lawless (1978), using a set of 12 diverse odorants, found that recognition for them at 28 days was about 76%, as did Lawless and Cain (1975), who, using 11 diverse odorants, also found that recognition was also about 76% at 28 days (see fig. 7.1). Although limitations are inherent in comparing across studies, all used a diverse range of odorants and all used a recognition memory procedure. The interesting finding here is that when the set size is larger (i.e., 21 vs. 11 or 12 odorants), hit rate drops, just as it did in the two short-term studies described above. Engen and Ross (1973; experiment 2) directly demonstrated that increasing set size decreased hit rate (by about 10% when the set size was halved) when participants were tested at three months—clearly a task in-

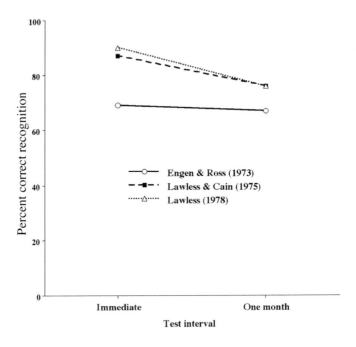

Fig. 7.1. Mean percent correct recognition scores for odors tested immediately and after approximately a one-month delay (data adapted from Engen and Ross 1973; Lawless and Cain 1975; Lawless 1978).

volving long-term memory. The conclusion suggested by these findings seems to be that set size can affect memory for odors at both short and long delays.

A second line of argument comes from the type of stimuli used in such experiments and suggests why capacity effects may appear. Jones et al. (1978) used similar smelling odors (herbs) and this seemingly had a debilitating effect on capacity, as recognition performance for a single odorant was approximately 72%, compared with about 82% for single-odor condition in Engen et al. (1973). Moreover, the more similar the members of the set, the more likely it is that they will interfere. Walk and Johns (1984) found that interpolating a distracter, similar to the target odor in a delayed matching to sample task, had the most deleterious effect on performance at short intervals (26 seconds).

Two further findings suggest that similarity also influences tasks involving longer delays, suggesting that both are sensitive to its effects. Rabin (1988) observed that unfamiliar odors were more poorly discriminated than familiar odors, suggesting that unfamiliar odors are judged as more alike, more similar. Perhaps not surprisingly, Rabin and Cain (1984) observed significant correlations between recognition memory performance at a 7-day interval and familiarity, suggesting that tasks likely to involve long-term memory are affected by set similarity. A more direct investigation was made by Engen and Ross (1973; experiment 2), in which either similar or dissimilar sets were experienced and with recognition then tested at three months. Recognition performance was some 13% worse when the stimuli were drawn from a similar set, rather than a dissimilar one. Although nobody has as yet directly contrasted short- and long-term tasks, both appear subject to capacity limitations by virtue of set similarity. There is certainly nothing as yet to suggest that qualitative differences emerge from the manipulation of delay.

That tasks at short and long delays are both affected by perceptual similarity directly relates to a further issue, coding strategy. At least two different forms of coding may be used in olfactory information processing—a perceptual code and a verbal code. To what extent, if any, does one dominate in tasks involving retention of olfactory information over a short period compared with longer periods? The evidence is ambiguous with respect to which mechanism of coding may be principally used, but it is not ambiguous in demonstrating that both types of coding occur at all delay intervals. First, there is clearly a perceptual component to tasks involving a short delay. Walk and Johns (1984) found that a similar odor, interpolated between the target presentation and the test phase, exerted a deleterious effect on recognition. Using an unrelated word, rather than an odor, had no effect, whereas a word semantically related to the target significantly enhanced performance above that of the control condition (no interpolated task). Perceptual effects on tasks involving short intervals have also been observed in other types of experiments. Jehl, Royet, and Holley (1994), found that paired comparisons, using either perceptually similar or dissimilar hard-to-name unfamiliar odorants, affected same/different judgments, primarily by increasing false alarms with pairs of different but qualitatively similar odorants. Although this effect only manifested when the interval between pairs was in excess of 10 seconds, a probable consequence of adaptation, it suggests that perceptual similarity does affect performance on tasks involving short delays. Similar findings have also been obtained by White et al. (1998). They approached this problem in a different way to try and assess the relative contribution of perceptual and verbal coding in a short-delay task. Their basic strategy was to use stimuli that were perceptually confusable (e.g., maple syrup vs. honey) and stimuli that were not, but that had similar names (e.g., lime vs. thyme). Both verbal and perceptual errors were observed, but perceptual errors were the most dominant.

Data from short-delay tasks suggest that both verbal and perceptual codes may be used. The same appears to be true when the delays are sufficiently long to infer a role for long-term memory. For example, Lyman and McDaniel (1986) observed that generating semantic information about the target odors enhanced subsequent recognition memory performance on test one-week later, although some doubt surrounds the replicability of this effect (Zucco 2003). A more direct approach has to been to exclude verbal coding through the use of various suppression tasks and then to assess how this affects recognition memory at both short and long (72 hours) delays (Annet and Leslie 1996). Annet and Leslie (1996) found that suppression had an equal effect on recognition memory at both delays, suggesting little difference in coding strategy between the two. It would appear that both verbal and perceptual coding strategies are used in both short- and long-delay tasks and that there is nothing to indicate the relative dominance of one strategy over the other at each type of interval.

A further approach that has been widely adopted in research on short-term memory, especially during the 1960s and 1970s, are serial position experiments. Here, recall of items from a list is better for items presented early (primacy) or late (recency). This was originally thought to reflect the operation of long-term and short-term memory processes, respectively (Atkinson and Shiffrin 1968). Although two papers have reported recency effects, with olfactory stimuli, in the absence of a primacy effect (Annet and Lorimer 1995; White and Treisman 1997), only one study has reported both recency and primacy effects (Reed 2000). Reed (2000) presented participants with a set of five unfamiliar hard-to-name odors and after the list was complete, then presented two more odors, one a foil and one drawn from one of the five positions in the list. Participants' task was to identify which had occurred before. Reed (2000) observed that identification was better when an item was drawn from either the early part of the list (primacy) or the later part of the list (recency). This effect was not altered if participants engaged in an articulatory suppression task, suggesting that the result did not arise from covert verbal labeling. These results have recently been criticized. Miles and Hodder (forthcoming) have been unable to obtain a primacy or recency effect in seven attempts to replicate Reed's (2000) procedure. At best, then, the presence of a recency effect for odors appears to resemble findings from other sensory modalities where difficult-to-identify stimuli are used and a recognition, rather than a recall strategy, is employed (e.g., Neath 1993). However, it is now acknowledged, in general, that primacy and recency effects probably do not reflect the operation of long- and short-term stores. First, primacy and recency effects can occur when the interval between items on a list are weeks or years (e.g., Baddeley and Hitch 1974). Second, primacy and recency effects, which occur when long intervals are used between list items, are affected by precisely the same variables that affect primacy and recency effects for lists with brief intervals between items (e.g., Greene 1986). This suggests that the two are not mediated by different processes, as some have argued. Instead, these results cast doubt on the usefulness of these types of data for telling us anything about dissociable long- and short-term stores.

A final source of evidence, which has been marshaled to support discrete short- and long-term stores, is the presence of neuropsychological dissociations, such that one patient demonstrates intact long-term memory with a specific deficit in short-term memory and vice versa in a second patient (c.f., Shallice 1988). It has been suggested that damage to the right temporal lobe, especially that resulting from epilepsy or following temporal lobe resection, may produce specific deficits in short-term olfactory memory (e.g., Carroll, Richardson, and Thompson 1993). Rausch, Serafetindes, and Crandall (1977) had patients with left and right anterior temporal lobectomies engage in a matching to sample task, in which they sniffed a target, followed 10 seconds later by four odors, one of which was the target. The participant's task was to identify the target, and those with right-sided damage were significantly poorer at doing so. Using a similar technique and population, Abraham and Mathai (1983) observed a similar right-sided deficit, as did Carroll et al. (1977) using a recognition memory procedure with an immediate test. In this case, however, right-sided deficits only appeared for easy-to-name odorants.

Although these findings seem to point to a specific right-sided, temporal lobe locus for short-term retention of olfactory material, two types of findings suggest that this interpretation is likely to be incorrect. First, three studies using immediate and delayed tests with a recognition memory procedure have identified deficits in *both* patients with left- and right-sided lesions, at *both* short and long delays (Eskenazi et al. 1983; Martinez et al. 1993; Dade et al. 2002). Second, Jones-Gotman and Zatorre (1993) tested left and right temporal lobectomy patients using a recognition memory test, immediately after exposure, after 20 minutes and after 24 hours. Although in this case a right-sided deficit was obtained, consistent with the reports identified above, the deficit was *consistent* across all three intervals. In summary, lesion location, left or right, has at best an inconsistent effect on performance. Most tellingly, when memory is tested at intervals likely to be indicative of short- and long-term memory, performance is uniformly bad at all intervals. It is unlikely that right-sided temporal lobectomy is associated with a specific deficit in retaining olfactory information over short periods. In fact, as we noted in a previous chapter, the only consistent neuropsychological dissociations that are observed are those between qualitative and quantitative olfactory discrimination (White 1998). In sum, there is no compelling evidence that there is a dissociable short-term store in olfaction. Information clearly *can* be retained for short or long periods, but there appears to be no consistent capacity, coding, or neuropsychological evidence, to suggest two qualitatively different stores.

Long-term Memory

Several studies have demonstrated that olfactory information can be retained for extended periods. The earliest such study was by Engen and Ross (1973). Participants smelled 48 odorants (experiment 1) and were then tested on 21 pairs (old vs. new), at intervals of up to 30 days. Not only was performance well above chance at 30 days, but hit rate across time was relatively flat, indicating little forgetting, albeit with rather poor initial recognition when compared with words. Similar findings have been obtained by Lawless and Cain (1975) and by Lawless (1978). The latter paper (Lawless 1978) is of especial interest as the author compared recognition memory for odorants with that for ambiguous shapes and pictures. Recognition memory for pictures was much better on the immediate test, but it decayed rapidly, relative to memory for odorants and ambiguous shapes, which were both poorly recognized initially but had much flatter decay slopes over the periods studied (up to 28 days).

At least three other studies indicate persistence of odor memory. Murphy et al. (1991; experiment 2) had young and old participants smell 20 odors and view pictures of 20 faces as well as 20 drawings of abstract machine symbols. Recognition memory was measured immediately, at 2–3 weeks, and at 6 months. Although for young participants correct recognition was still above chance for the odors at 6 months, performance was at chance level for the elderly by 2–3 weeks. Performance for faces was comparable to odors in younger participants, but better in older participants. The drawings were recollected better at all intervals by both groups. Goldman and Seamon (1992) used a very different approach to explore retention by having participants attempt to name odors they had probably not smelled since childhood. They found that college students were able to identify such odors at significantly better than chance, suggesting retention of olfactory information over several years. Finally, Haller et al. (1999) reported that 28 years (on average) after exposure to vanillin in baby milk, bottle-fed adult German participants preferred the flavor of vanillin-adulterated tomato ketchup to the unadulterated form, when compared with breast-fed participants. This finding suggests the retention, in this case of an odor preference, for most of the participant's life. Clearly, odor memory can be very long lived indeed.

Directly related to studies of the longevity of odor memory are those exploring its susceptibility to interference. The most widely cited example of resistance to interference (over 100 citations according to the Institute for Scientific Information [ISI]) is provided by Lawless and Engen (1977). However, we have chosen to discuss this in the next section, as it may not strictly reflect resistance to interference of long-term odor memory per se, because the study itself explored resistance to interference of effortfully acquired odor-picture associations. Rather, in this section we stick to exploring interference and its effects that are either incidental (i.e., naturalistic) or that pertain solely to olfaction. In this respect there are only two studies that are directly relevant, apart of course from those that we discussed earlier in chapter 6. The first of these utilized a rather unusual correlational approach (Koster, Degel, and Piper 2002). Participants were covertly exposed either to a different odorant in each of two visually distinctive rooms, the same odor, or to no odor(s) at all. Later, participants were asked to judge how well a set of odorants, which included the two target odorants, fitted pictures of a variety of different scenes (e.g., a bank, railway station, etc.), including the two target rooms. They found that when a participant had, for example, experienced lavender in one room and then lavender in the second room as well, there was evidence of interference, in that the most recent experience was judged a better fit than the earlier experience. However, this effect only occurred in participants who could not verbally label the two target odorants. Participants who could identify the odorants did not show any evidence of learning associations between the odors and the rooms, so no interference could be expected. Although these findings are based on a post hoc split into identifiers and nonidentifiers, which must raise a concern over whether other factors were responsible for the effect, these data are of interest as they contradict the widely held assumption that odor memory is especially resistant to interference. In this case, at least, this does not appear to be correct.

The second study by Zucco (2003; experiment 2) stands in marked contrast. Zucco utilized an experimental design in which participants were split into three groups. The easiest way to describe the design is from the perspective of the group that received the olfactory stimuli. On their first session they were asked to sniff 15 odors, after which they were given a recognition memory test. There were then three subsequent sessions, each separated by one week. These included (1) a condition in which 15 new odorous stimuli were smelled, followed by 15 visual stimuli, followed by an odor recognition memory test; (2) 15 new odor stimuli, followed by 15 auditory stimuli, followed by an odor recognition test (i.e., both conditions 1 and 2 represent cross-modal interference conditions); and finally (3) a within-modality interference condition, in which 15 new odor stimuli were followed by 15 further olfactory stimuli, then followed by an odor recognition memory test for the older set. The other between-group factors in the design were a visual and an auditory condition (i.e., as above, but visual or auditory stimuli were experienced first). Zucco (2003) found that olfactory interference had no effect on odor recognition memory, but that auditory interference affected recognition of auditory stimuli and visual interference affected recognition of visual stimuli. All modalities were unaffected by cross-modal interference. In this experiment, at least, olfactory memory, when tested shortly after the interference phase, was resistant to interference. One important caveat here is whether these results are only applicable to short-term memory processes, given that the interval between exposure and testing was quite brief (i.e., the time needed to experience the interference stimuli). However, as we noted in the preceding section, there are no compelling grounds to suspect that different memory types exist based on task interval.

Several studies have explored whether enhancing encoding, by the use of verbal labels for example, can improve subsequent long-term recognition memory. These studies are of interest for two reasons. First, as with the data on short-term memory, they have something to tell us about the way in which olfactory information is encoded and stored for long periods, that is, to what extent does this rely on, for example, a verbal or a perceptual code, or both? Second, as we discussed earlier, odor discrimination is known to improve when participants can verbally label the odorant. This effect could result from the additional information that the verbal label brings to the testing situation, something that may be of considerable importance given that all olfactory discrimination tasks involve serial presentation and require participants to retain information over an interval. A further possibility, and one that is not mutually exclusive, is that odor naming has a more specific effect on perception, that is, the generation of an odorant's name may act to refine the percept in a number of possible ways as discussed in the introduction to this chapter.

Although two studies have reported failures to obtain enhanced recognition memory using verbal labels (Lawless and Cain 1975; Zucco 2003), a far larger number of studies have obtained positive results. Engen and Ross (1973; experiment 3) had participants smell 20 odorants. Participants were either provided with the odorant's name or with associations commonly provided by other experimental participants when the odorant had been smelled before. A recognition test three months after this training phase revealed a small but significant improvement in recognition in the name group (76% vs. 70% correct). Using an individual differences approach, Rabin and Cain (1984) examined recognition memory for 20 target odors after a seven-day interval. Odorants that could be correctly named (and consistently so) were better recognized than those that could not.

The most-oft-cited studies with respect to the effect of encoding strategy on recognition are those of Lyman and McDaniel (1986, 1990). Their 1986 study examined the effect of four encoding strategies; mere exposure to 30 odors, being asked to form a visual image of the odor's source, generating a life episode related to the odor, and, finally, generating a name for the odor. Seven days later participants were given a recognition memory test. Recognition memory performance, as assessed by d' was best in the name group and poorest in the mere exposure group. Hit rate did not significantly differ by group, however; the effect manifested through a reduction in the false-alarm rate. A similar approach was adopted in Lyman and McDaniel's 1990 study, in which the same odorant set was employed, but the strategies differed in that, in their study, participants in the enhancement groups were provided with information instead of having to generate it themselves. The groups were again a mere exposure control, a picture group who received an illustration of the odorant's source, a verbal label group, and a combined picture and verbal label group. Recognition was tested seven days later and was best in the combined group, intermediate in the label- and picture-alone groups, and poorest in the exposure condition. Several more recent studies have confirmed the general impression left by these studies that providing additional information during encoding, especially a verbal label, enhances subsequent recognition memory performance (Murphy, Cain, Gilmore, and Skinner 1991; Cain and Potts 1996; Jehl, Royet, and Holley 1997; Larsson and Backman 1997).

As we noted earlier, it is of some interest to know exactly how naming, in particular, may enhance performance on recognition memory tasks. One possibility is that identification and the generation of a name actively affects perception. In this respect, the Cain and Potts (1996; experiment 2) findings are suggestive. They asked participants to name 20 odorants. Two days later they returned to the laboratory and completed a recognition memory test. The interesting part of this experiment concerned misidentifications during the initial naming period. During the two-day interval, the misidentified items were replaced by the actual misidentified odorant, which was then used in the subsequent recognition memory phase on a participant-by-participant basis. For example, if a participant had smelled ginger and reported it as garlic, then garlic odorant, rather than ginger, was employed in the recognition memory test. Participants often fell for these baits, suggesting that in the example above, when smelling ginger and reporting that they had smelled garlic, what they *may* have been experiencing, perceptually, was garlic too (see Engen [1987] for a similar explanation of odor misidentifications that occur during naming experiments).

Paired-Associate Learning

Studies of olfactory paired-associate learning can be broadly divided into two categories, those examining whether it is different from paired-associate learning in other sensory modalities and second, those that solely examine the properties of olfactory paired-associate learning. We deal with each of these approaches separately. It was thought for many years that paired-associate learning between odorants and some other cue (e.g., numerals, pictures, colors) is less efficient than paired-associate learning between other forms of stimuli (e.g., geometric shapes, colors) and the same cue (e.g., Heywood, Vortried, and Washburn 1905; Bolger and Titchener 1907). However, it is only more recently that the similarities and differences have been explored in more depth. Two sets of investigations by Davis (1975, 1977) are important in this respect. Davis (1975) demonstrated that paired-associate learning between free-form shapes and numbers and odorants and numbers differed, in that the latter were acquired more slowly. Davis (1977) examined whether this difference merely reflected characteristics of the stimulus set, most notably intraset similarity, rather than characteristics of the modality in which the set was perceived—a potentially important distinction. Davis (1977) showed that even with sets equated for similarity, the acquisition of odorant-number associations was slower and resulted in a lower final performance level than that with free-form shapes.

More recent studies have pointed to a somewhat different conclusion. Bower (1994) had participants learn associations between either: three color-color pairs, three odorant-odorant pairs, three odorant-color pairs, or three color-odorant pairs. Participants in each one of these four groups received a total of 15 pairings and unlike Davis (1977), Bower et al. (1994) found that learning involving odors was as good for color-color pairs *if* the odorant came after the color. However, with odorant-odorant pairs and odorant-color pairs (i.e., the same order as for Davis [1977]), learning was significantly poorer at the end of training, suggesting that cue order has a major impact on paired-associate learning with odorants but not colors. Bower et al. (1994) suggested that this reflected the ecology of odor perception, in that odorants are normally associated with the sources from which they emanate (i.e., cue then odor); that is, we identify the source and then associate it to the odorant. Alternatively, a further way to account for this finding would be that, where an odorant comes first, resources are devoted to generating its name at the cost of paying attention to the stimulus that follows.

Three studies, all conducted by Herz and colleagues, have suggested that odor paired-associate learning is as good as that between stimuli from other modalities with the same cue. In all these experiments, as with Bower (1994) above, the actual number of stimulus pairs to be acquired was more limited (three pairs) when contrasted with Davis's two studies, both of which used larger sets (six pairs). Davis (1977) himself observed that greater within-set similarity retarded acquisition and so at least some of the advantage seen in Herz's studies may be attributable to this cause. A further possibility is that in all the Herz studies (i.e., Herz and Cupchik 1995; Herz 1998, 2000), the to-be-paired stimuli were far more emotive (pictures) than the cues used in earlier studies. Consequently, it seems that the task itself was far more engaging, even though learning was incidental. In sum, although more recent studies demonstrate that acquisition may not always be poorer in olfaction, these may reflect the most optimal conditions. Where set size is larger, the odor cue comes first and the task is less engaging, differences will likely emerge in the ability to engage in paired-associate learning with odors.

The second category of paired-associate learning study is one in which there is no nonolfactory control condition. Such studies have concentrated almost exclusively on the acquisition of verbal labels. This focus arises naturally from the observation that most participants have great difficulty in generating names for even the most commonly encountered odorants when appropriate contextual cues are absent (Engen and Pfaffman 1960; Engen 1987). The first such study to explore paired-associate learning in this context was conducted by Cain (1979), who outlined three hypotheses as to why odor naming might be so poor; sluggish formation of associations, impoverished retrieval, and inherent confusability of olfactory stimuli. The last mentioned of these formed the motivation for many of the early studies on odor naming (Engen and Pfaffman 1960; Desor and Beauchamp 1974).

Using only female participants, Cain (1979) found that presenting feedback about the correctness, or not, of self-generated labels improved performance from 48/80 odors correctly identified to 62/80. More impressively, when a further group of female participants were provided with a label by the experimenter and given corrective feedback, performance increased from 62/80 (after the first trial with information presented) to 75/80 by the third trial. Although these experiments do not readily determine which of the three factors identified by Cain (1979) are most important, they do suggest that inherent confusability must be the least relevant, because how else could one obtain such good terminal performance (i.e., 75/80) if the odorants were inherently confusable?

Other studies too have found that training will improve the ability to generate an odor's name. Cain (1982) reported that both men and women benefited almost equally from training with either self-generated labels (plus feedback) or experimenter-generated labels (plus feedback). With 40 odorants, performance improved by an average of about 6% over five trials for self-generated labels and 12% for experimenter-generated labels, and most of this difference reflected the greater accuracy of the experimenter-provided labels. Desor and Beauchamp (1974) also explored whether training could improve nameability. They too, albeit with only three participants, found that almost perfect performance could be obtained for a large set (64 odorants) after several trials with feedback.

A consistent finding in the literature has been the advantage that female participants have in naming odors when compared with men (e.g., Gilbert and Wysocki 1987; Cain 1982; Engen 1987). Several possibilities emerge as to the cause of this difference, the most basic being that it stems either from an inherent biological difference between men and women, or experiential differences, namely in the greater attention given to odorants by women than men (Herz and Cahill 1997). Dempsey and Stevenson (2002) explored which of these two alternatives might be correct by having male and female participants learn associations between unfamiliar odorants and novel Swahili names, thus ruling out any a priori benefits women might have in identifying these particular stimuli. Both men and women acquired the associations equally well and at the same rate, suggesting no differences in attention or motivation by gender. However, when they were retested one week later, male participants were poorer at retrieving the odor names relative to female participants. A second experiment replicated this finding and eliminated the possibility that it might reflect differences in sensitivity to interference (i.e., men being more sensitive). The results suggest a biological origin for differences in naming ability, a similar conclusion to that reached by a recent review on this topic (Brand and Millot 2002).

As we noted earlier, olfactory memories can be retained for long periods and there is some evidence that they are also especially resistant to interference (Zucco 2003). This raises the question of whether associations to odors are also similarly resistant to interference. Although one experiment presented earlier, using incidental learning, casts some doubt on this (Koster, Degel, and Piper 2002), two findings described below suggest that it is. Note, however, that neither of the experiments reported below had a cross-modal control condition, so there must be some concern as to whether the failure to find retroactive interference resulted from an insensitive experiment rather than some special characteristic of olfactory paired-associate learning.

Lawless and Engen (1977) had participants learn associations between odorants and picture postcards, with the postcards always available for inspection during learning and testing. Although olfactory paired-associate learning was susceptible to proactive interference, that is, the second learned of two sets was more poorly retained, it was not susceptible to retroactive interference, that is, the first learned set was not affected by subsequent learning of a new set. Similarly, Dempsey and Stevenson (2002) had participants learn new word associations to the same set of odorants. They found, as did Lawless and Engen (1977), that this new learning did not affect retention of the previously learned material. Both of these studies indicate that once an odor association is acquired, it is apparently difficult to alter subsequently.

Retrieval Cues

Several experiments have shown that the incidental presence of an odorant during a study phase can later enhance retrieval if it is present during the test phase. Cann and Ross (1989) provided an early demonstration of this effect by passively exposing male participants to fifty female faces. While viewing the faces one group of participants smelled a pleasant odor, a second group smelled an unpleasant odor, and a third group smelled no odor at all. Forty-eight hours later participants took part in a face recognition procedure utilizing fifty new female faces and fifty old ones drawn from the original set. The odor present during this task was either the same as during the study phase, different, or, again, no odor was present at all. Discrimination of old and new faces as measured by d' was significantly better when the odors at study and test were the same, compared with if no odor was present at test/study or if the odorants were changed. The effect did not primarily arise through changes in hit rate but through a reduction in false-alarm rate.

Not only can odors successfully act as retrieval cues with faces, they can also assist in recalling the members of an incidentally studied word list. Schab (1990), in two studies (experiments 1 and 2), found that the presence of an odorant on study and test significantly improved recall accuracy by about 10%. In his third experiment, participants received an incidental word-learning task in the presence of one of two odors. On test, the odor was either the same or different from the study phase. All participants then completed both an explicit recall task and an implicit stem-completion task. When the same odor was present on study and test, performance was significantly improved over when different odors were present on the two occasions. However, although this effect was detected for the explicit recall task, there was no effect on the implicit stem-completion task.

A more recent study investigated whether odors might also serve as retrieval cues for other types of task such as those involving procedural memory. In their first experiment Parker, Ngo, and Cassaday (2001) demonstrated that they could obtain an effect similar to Schab's (1990) experiment 3 (see also Pointer and Bond [1998] for a further demonstration). In their second experiment, participants completed three types of assignment; they learned a word list, completed the Tower of Hanoi puzzle (in two forms), and the Porteus Maze. These were completed with either a lemon or lavender odor suffused into the room. The next day participants returned to the same room, which was then either suffused with the same odorant or a different one from the study phase. Participants attempted to recall the words learned the day before and completed the Tower of Hanoi and the Porteus Maze again. Although both performance on the word list and the Porteus Maze was enhanced when test and study odorants were the same, compared with the different condition, performance on the Tower of Hanoi only improved in the presence of lavender odor, suggesting a specific effect of this particular odorant.

Evidence for such odor-specific effects is rather mixed. Although Baron and Bronfen (1994) found odor-specific enhancement effects on a word construction task, Knasko (1993) could obtain no evidence for such effects on a range of verbal and mathematical problems. Knasko (1993) later asked participants what effects *they thought* might occur with the pleasant and unpleasant odorants used in her study. Participants' expectations were that they would affect mood and performance—contrary to what was actually observed, but raising the question of whether odor-specific effects in other studies might in part derive from participant expectation. Although odorant-specific effects do occur (Baron and Bronfen 1994; Knasko 1995), odorants can clearly serve as retrieval cues and such effects are generally not odor specific but context specific—the same odor context has to be present on study and test.

Several experiments have demonstrated that olfactory cues can serve to retrieve autobiographical memories—that is, personally relevant episodes from one's life. A particular memory, when retrieved with an olfactory cue, appears both more emotionally laden than when retrieved via other modalities and may refer back to more temporally remote incidents too. Given this opening statement, it is perhaps surprising that the first formal study failed to find such effects (Rubin, Groth, and Goldsmith 1984). This may have been because it did not control for the type of specific memory elicited by cues from different sensory modalities. Instead, later studies have examined the effect of different modality-cued recall on the *same* memory, and it is largely with this technique that the emotional enhancement effect of using an olfactory retrieval cue has been obtained.

Taking first the issue of greater emotionality, Herz and Schooler (2002) utilized a two-stage recollection procedure, in which participants first recovered an autobiographical memory to a word and then attempted to recover/elaborate the same memory utilizing either a visual or an olfactory cue. This technique did obtain evidence that odor-induced autobiographical memories were reported as being more emotive. Herz (2004) recently extended this technique by again cueing a memory using a word (e.g., cut grass) and then having participants recover the same memory by using an olfactory cue (e.g., *cis*-3-hexanol), an auditory cue (e.g., the sound of a lawnmower), and a visual cue (e.g., picture of a lawn being cut). Yet again, olfactory cues evoked a more emotionally potent memory, than that evoked by the visual or auditory cues, which did not differ.

That olfactory information can assist retrieval of information encoded a long time ago has been shown in two studies (see the Goldman and Seamon [1992] study cited earlier as well). The first by Aggleton and Waskett (1999) obtained a sample of participants who had visited the Jorvik Viking Centre (York, UK) about six years previously. These participants were then formed into three groups. One group smelled a series of odors that were specifically used at the center to enhance the Viking experience and which they likely experienced while there. During the process of sniffing these odors they completed a questionnaire about the center (i.e., what exhibits were there, etc.). They then completed the same questionnaire again, but this time while

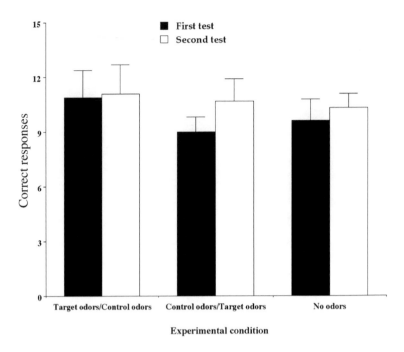

Fig. 7.2. Mean correct responses for participants who smelled the target odors on the first test and the control odors on the second, for participants who did the reverse, and for those not receiving any odors (data adapted from Aggleton and Waskett 1999).

smelling a nonrelevant control set of odorants. A second group completed the same two tasks, but in reverse order (i.e., irrelevant odors first, then relevant odors). A third group completed the questionnaire twice, but did not have any odors present on either occasion. Although there was a tendency for better recall of information when the Viking relevant odors were present on the first questionnaire completion, the most marked effect (see fig. 7.2) was when the second group completed the same questionnaire now with the Viking relevant odors present. They showed a significant and marked improvement in performance, whereas the nonodor group was poor relative to both of these groups, even on the second presentation. Although it is arguable whether the material recalled may strictly fall within the definition of personally relevant information, participants clearly recalled more when the Viking odors were present.

Research on autobiographical memory has revealed what is termed a retrieval bump, in that, if personal memories are recovered from participants, most will be either about their current lives or about their early life as a teenager or young adult (e.g., Conway and Pleydell-Pearce 2000). A similar finding was obtained by Chu and Downes (2000) when they too used words as the retrieval cue, but when they used odorants equivalent to the words (e.g., the word grass and the odor associated with grass) they obtained a different distribution of memories, with more clustered around childhood. That is the reminiscence bump was displaced further back in time, suggesting that odors may serve as cues to access memories that are not so readily accessed by other means.

Priming

One dominant area of memory research during the past forty years has been priming, in which incidental exposure to some cue later results in a change in performance (for good or bad) on some subsequent task. Several researchers have attempted to obtain similar findings in olfaction, although, as we detail below, this work has been disappointing in general and the results do not as yet yield a coherent picture. Schab and Crowder (1995) reported some of the earliest work on olfactory priming. They found that presentation of an odor and name, or presentation of an empty bottle with an odor name (participants being forewarned that some odors would be very faint), enhanced naming for these odorants on a subsequent recognition task. Enhancement of identification performance was most pronounced for the real odor and name condition (72% correct), compared with the name-only condition (53% correct) and the control condition (32% correct). Similar findings were also obtained with naming latency. Two caveats need to be born in mind about these findings. First, as Schab and Crowder (1995) acknowledge, this effect could be mediated by semantic rather than solely perceptual processes. Second, they failed to find a similar result in a conceptual replication. Moreover, they also failed to obtain any evidence of priming in tasks which might have been expected to be more perceptually based, namely detecting odorants at or near threshold following exposure to the stimulus and its name, or the name alone.

In a recent review of the olfactory priming literature, Olsson, Faxbrink, and Jonsson (2002) described several priming experiments conducted by a German group (Wippich and colleagues 1990 and 1993, as cited in Olson et al. 2002) and published in a German psychology journal—thus not widely available to an English speaking audience. Both studies found some evidence of priming, although again the effects were neither robust nor always consistent with the experimenter's expectations. First, they found that exposure to an odor, but not to its name, enhanced subsequent attempts at identification. Second, they observed that exposure to pairs of odors, but not to the names, enhanced later judgments of whether these odor pairs were the same or different (with no effect on novel control pairs). However, this effect did not occur when participants had been exposed to single odorants that made up the pairs, making it difficult to draw the most obvious analogy to the exposure and discrimination studies discussed in chapter 5.

The most extensive series of reports on odor priming come from Olsson and colleagues. Olsson and Cain (1995) found that preexposure to an odor enhanced identification latencies for that odor, but only when the test was conducted via the left nostril—no such effect was obtained for presentation to the right nostril. Olsson (1999) reported another priming type effect, in which following odor preexposure, judgments of same versus different for a preexposed target against comparison odorants was faster, but only if the target could not be named. Finally, Olsson and Friden (2001) found that edibility judgments were made more quickly for a preexposed set of odorants than for nonpreexposed odorants but that these judgments were no more accurate. Although all this work on priming suggests that *something* happens, clearly the picture that emerges is neither simple, nor is it yet complete.

Imagery

As we described in the opening of this chapter, imagery may play an important functional role in olfaction. First, as we have already seen, naming an odorant can result in an enhancement of its discriminability from other odorants (Rabin 1988). One possibility alluded to earlier is that this effect results from the generation of an image, one possibly at variance with that generated directly by the stimulus. The process might work something like this—when an odorant is smelled many olfactory object encodings are activated (e.g., A, B, and C), a name is retrieved (e.g., A), followed by backward activation of the olfactory memory associated with A or deactivation of those not associated with A. Whatever the precise mechanism, this process would reflect a highly entwined relationship between stimulus and mnemonic processes and have, on some occasions at least, the practical consequence of enhancing the discriminability of the olfactory percept.

A second function of imagery relates back to the earlier discussion of the multisensory aspects of olfactory perception. Clearly, when a participant smells caramel odor, for example, and reports that it smells sweet, for reasons we discussed earlier it is quite appropriate to describe the experience of sweet as an image—it is certainly not a direct consequence of the odorant's physiochemical makeup, but an effect of memory. In this case its functional significance would be to provide extra information about the stimulus through a purely perceptual channel (i.e., this tasted sweet before). When a multisensory cue is available while smelling an odor, such as color, for example, a different type of effect may occur in which the cue directly activates an odor object memory. The effect on odor perception may be profound, resulting in a total revision of the bottom-up generated representation so as to minimize perceptual dissonance between the dominant sense (vision) and the minor sense (olfaction). One possible example of this is the effect of coloring white wines red, as discussed elsewhere in the text.

The third functional role of imagery is in olfactory thought. That is, can we summon up an olfactory representation on demand when there are no chemical cues available and utilize this information productively? Unfortunately, the precise functional benefit of volitional olfactory imagery has been rarely discussed (but see Cain and Algom 1997) let alone examined, but we might expect that perfumers, flavorists, and chefs, for example, might engage in something like it when they design novel flavor or odor combinations. To date, the evidence for volitional imagery has been equivocal and many leading researchers in the field have tended to either dismiss our capacity for it (e.g., Crowder and Schab 1995; Engen 1991; Herz 2000) or robustly support it (e.g., Cain and Algom 1997; Elmes 1998). In this section we briefly review the evidence—for a more detailed account the reader is advised to see two recent articles that consider this issue in greater depth (Elmes 1998; Stevenson and Case 2005).

The first question we might consider is a very broad one indeed: Can people *ever* experience an olfactory sensation when a chemical stimulus is absent? We might view favorable evidence here as an existence of proof, thereby excluding the possibility that the human brain is incapable of forming an olfactory image without a chemical stimulus. This approach demands the study of phenomenology, as it directly pertains to participants' mental experience; thus we have to rely on self-report. Some of the strongest phenomenological evidence favoring imagery comes from the study of olfactory hallucinations, especially those which occur in conditions that are not typically associated with significant psychopathology (notably delusional states), such as epilepsy, migraine, cerebral aneurysm, and posttraumatic stress disorder. In each of these conditions we can find evidence of people reporting olfactory hallucinations and spontaneously behaving in a manner entirely consistent with that hallucination, perhaps the gold standard of self-report data. A few examples should suffice to illustrate this.

Reports of olfactory auras (hallucinations) are quite common in epilepsy and there are several reports of participants behaving as if they had smelled something, when in fact they were experiencing an aura. Efron (1956) reported the case of a woman who, while picking flowers, experienced what she thought was their smell. She asked her friend to verify the smell. Her friend reported that the flowers had no odor. Daly (1958) described two further cases, one who tried to open a window to get rid of a disgusting smell and another who smelled a strong peach odor, but whose friend was totally unable to smell it. Embril et al. (1983) and Scully, Galdabrini, and McNeely (1979) separately identified two similar reports, in which the person experienced an odor so repellent, that they both had their houses extensively searched in an attempt to locate its origin.

Migraine is not a condition one normally associates with abnormal behavior, yet olfactory hallucinations are not uncommon and again instances can be found of contextually appropriate behavior. For example, Crosley and Dhamoon (1983) had a patient who experienced the smell of gas prior to migraine onset and their home was frequently investigated for gas leaks. Like migraine, cerebral aneurysm can also produce olfactory hallucinations with minimal psychopathology. Toone (1978) described a patient who on walking into a hotel was engulfed by the aroma of a delicious roast dinner. On complementing the landlord on his culinary skills he was told that no food was or had been prepared, and that there was no such smell either. Finally, a completely different condition, posttraumatic stress disorder, can also result in the experience of olfactory hallucinations. Burstein (1987) described the case of woman who had been in an automobile accident. After recovering from the physical effects of the accident, she was traveling in a car with her husband when she was overcome by the smell of petrol—a smell that her partner could not discern, but which was intimately associated with her accident.

Although these examples suggest that people can experience olfactory sensation in the absence of appropriate stimulation, one would still wish for evidence derived solely from healthy participants. Here the picture is equally interesting, but for very different reasons. A century of research on volitional olfactory imagery has reached one definite conclusion, most participants find it very hard to imagine odors (e.g., Betts 1909; Gilbert, Crouch, and Kemp 1998; Stevenson and Case, forthcoming). Even if they can imagine them, the resulting image is typically judged as less vivid and clear than images evoked from all of the other sensory modalities (e.g., Ashton and White 1980). Although such self-reports do suggest that some participants can evoke olfactory images, many participants clearly have great difficulty in doing so.

Reliance on self-report data alone can never provide us with closure on the existence, or not, of imagery. In response to this a series of highly inventive approaches have attempted to test whether imagining an odorant results in performance on a subsequent task similar to that generated by the real thing. The underlying premise here is that imagery invokes the same (or similar) processes to those used in perception, consequently imagining an odor should result in similar task performance when compared with a real odor control group. It needs to be stressed here that we can only infer that this involves a mental image. For example, similar processes could be evoked without any conscious sensation. More importantly, similar outcomes could be generated by reliance on explicit knowledge (Schifferstein 1997), verbal coding (Herz 2000), or in general, rather than specific, image-based effects (Segal and Fusella 1971). All these possible causes need to be carefully evaluated when deciding whether a particular experimental outcome does indeed support the existence of olfactory imagery.

Two general findings emerge from this experimental approach to imagery. First, bar one exception (Lyman and McDaniel 1990), all studies demonstrating parallel performance between an imagery and a real odorant condition, have used a pretraining phase in which participants learn to associate the cue (e.g., a word) with the odorant which is to be evoked (i.e., Algom and Cain 1991; Algom, Marks, and Cain 1993; Stevenson and Prescott 1997; Djordjevic, Zatorre, and Jones-Gotman 2004; Djordjevic, Zatorre, Petrides, and Jones-Gotman 2004). Second, that even with the studies that have demonstrated parallel performance-memory psychophysics (e.g., Algom and Cain 1991; Algom, Marks, and Cain 1993; Stevenson and Prescott 1997), recognition memory (Lyman and McDaniel 1990) and threshold effects (Djordjevic, Zatorre, and Jones-Gotman 2004)—only one study to date has managed to eliminate all of the interpretive problems alluded to above. This is not to say, of course, that the others do not demonstrate an imagery effect, but it is simply that we cannot be sure. The most compelling illustration is that provided by Djordjevic, Zatorre, Petrides, and Jones-Gotman (2004).

Djordjevic, Zatorre, Petrides, and Jones-Gotman (2004) first assessed participants' threshold for lemon and rose odor. Participants were then assigned to one of three conditions: olfactory imagery, visual imagery, or a no imagery control. The olfactory imagery group repeatedly smelled lemon and rose odor (with names available) until they could satisfactorily evoke an image, as did the visual imagery participants, but with photographs rather than odor-ants. All participants then completed forced-choice detection trials, in which they received their threshold level for the target odorant versus a water blank. In the two imagery conditions, participants were asked to imagine either the smell/sight of lemon or rose. These imaginings were consistent with the target odorant on half the trials and inconsistent on the other half. Control participants received the same experimental task but did not receive any cues. The principal finding from this study was that odor imagery, but not visual imagery, significantly interfered with detection accuracy, but did not facilitate it, relative to the control condition (see fig. 7.3). These results are particularly convincing because (1) they are not based on a null result; (2) they exclude any generalized imagery effect because if it were present, the visual imagery condition would reveal it; and (3) verbal mediation is unlikely because the visual condition also received verbal instructions to image, but without any measurable effect on performance. The basis for this effect in mental imagery is made stronger by the finding of a correlation between per-

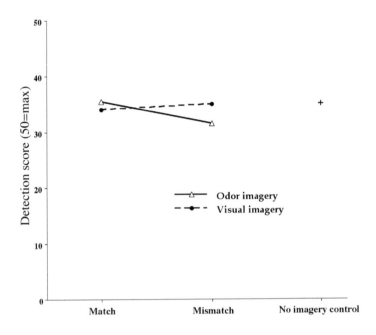

Fig. 7.3. Detection accuracy at threshold under conditions where the imagined stimulus either matches or mismatches the target and for a nonimagery control condition (data adapted from Djordjevic et al. 2004).

formance on this experimental procedure and self-reported olfactory imagery ability.

One source of evidence for the perceptual basis of visual imagery has been the observation that brain areas involved in forming a visual image are similar to the brain areas involved in parallel perceptual activities (see Kosslyn and Thompson 2003). Although similar studies are in their infancy for olfaction, functional magnetic resonance imaging results support the notion that brain areas activated when smelling are the same as those activated when attempting to imagine a smell (Levy et al. 1999). The Levy et al. (1999) study also revealed that anosmic participants could also activate the same regions when attempting to imagine an odorant, whereas a second study revealed that congenital anosmics, namely those with no extant odor memories, could not imagine odors nor did they show an appropriate pattern of activation (Henkin and Levy 2002). The import of these findings is twofold. First, similar brain regions are activated during olfactory perception and olfactory imagery, although this could be an artifact of sniffing, a strategy commonly employed by participants when attempting to imagine a smell (Bensafi et al. 2003). Second, participants who do not have access to odor object memories—the congenital anosmics—could not activate olfactory brain areas during imagery, unlike acquired anosmics who would be expected to have intact odor object memories. In sum, people do seem to be able to experience imagery, but by and large they are not very good at it, possibly because they have difficulty in accessing perceptual level representations. They do, however, appear to get better at this if they are allowed to practice.

Olfactory Information Processing

This section attempts to integrate, albeit at a preliminary level, human olfactory information processing. The first step in this task is to outline what needs to be accounted for by any model. The answer to this is provided by the content of this chapter—a short- and long-term memory capacity, the ability to form associations with other events and to retrieve such information, and the capacity for imagery. All these requirements need to be tempered by what we know about olfaction's unusual characteristics—no clear dissociation between a short- and long-term store, impoverished ability to name odors, slow forgetting in recognition memory tasks, sluggish formation of new associations and their possible tenacity, and the apparently limited ability of most participants to evoke an olfactory image. In addition, we must be able to predict how the resultant model might handle priming, which we know relatively little about, and, most importantly, how this all relates to the mnemonic-based object recognition system which we believe underpins olfactory perception.

We start by describing a putative human olfactory object recognition model similar to that described in more depth by Stevenson and Boakes (2003). This model is envisaged to have one central component, a long-term store of olfactory objects—hereafter, the object store. The contents of the object store are the products of previously encoded outputs from the receptor-processing system, namely, the system that turns chemical stimuli into their initial neural representation. When a target odorant is smelled in its environmental context (i.e., with other odorants present too), the output from the receptor-processing system contains all this information, which then flows in parallel through the object store. If the target odorant has been smelled before, it will activate encodings that are similar to its part of the receptor processor output, enabling the selection of relevant information, the target odor, from background chemosensory noise. This selection process has three characteristics. First, the more similar the pattern from a target odorant to one already in the object store, the greater the activation of that object encoding. Second, an input pattern from the receptor-processing system can activate multiple object encodings, based on the first principle. Third, activation is stochastic, that is, the same input pattern may produce different outputs, albeit generally similar, on different occasions. These activations are presumed, in this model, to represent our experience of odor quality by virtue of redolence, similarity to previously encountered smells.

If an odorant is a novel combination, then it will result in activation of many encodings, but each will be activated only to a small degree. That is, in terms of similarity, the activation pattern resembles a lot of object encodings, but none to any great extent. The result of this will be a vague representation, one redolent of many other odor objects. Consequently, it will be hard to discriminate it from other unfamiliar odors, just as observed (e.g., Rabin 1988). If there is a familiar component in the output of the receptor-processing system, it will also activate many encodings, but a small proportion of these will be activated to a far greater extent, resulting in a more definite representation and better discriminability. In this case, the high level of activation of one or more encodings may prevent the object store from encoding the receptor-processing output. However, when activation levels are not uniformly high, as with the aforementioned example of an unfamiliar odor, then the store will encode the receptor-processing output as a new odor object.

The preceding text describes the perception of odor quality, but it does not establish how we experience the intensity and hedonic aspects of sensation. These too can be accommodated but in rather different ways. The intensity aspect of sensation has been dealt with before by considering it as a direct correlate of the level of activity in the receptor-processing system—more activity equates with more intense sensation, less activity with less intense sensation (see Lansky and Rospar 1993). There are three findings that indicate that this basic account requires some modification. First, as the case of HM shows (Eichenbaum et al. 1983), it is possible to perform successfully on an intensity discrimination task in the apparent absence of an ability to perceive odor quality. Of course this is perfectly consistent with the idea of two modules (receptor-processing system and object store), but it implies that the intensity processor must have some capacity to retain information for short periods. That is, this module must have some memory capacity, unless verbal labels are used instead. Second, the finding of sniff vigor constancy (Teghtsoonian, Teghtsoonian, Berglund, and Berglund 1978), in which changes in sniff size have relatively small effects on judged intensity of a particular stimulus, implies an imperfect correlation between sensation and receptor activity. That is, a vigorous sniff should increase receptor activity by providing more molecules to the receptor surface. This could be accommodated by including a feedback loop, because structures that control the musculature responsible for a vigorous sniff could suppress activity in the receptor-processing module producing constancy. Third, familiar odors are typically judged as smelling more intense than unfamiliar odors (Hudson 2002). Yet again, this implies an imperfect correlation between net receptor activity and sensation. Because the model we have advanced here suggests that the locus of familiarity is in the object store, the object store itself may add some component of sensation, which is additive to the intensive sensation generated by the receptor-processing system.

The hedonic aspects of olfactory sensation can be conceived in several ways, but whichever way is chosen, the process must reflect the apparent ease with which odors are able to evoke affect. One way to conceptualize this is a gating mechanism that is directly attached to each odor object in the store and so depends on the integrity of the object store. In this scheme, once an object has been identified, that is, there is a high level of activation for one or more encoding(s), each opens a gate to either a positive or a negative emotional affect system. This choice depends on the consequences that ensued upon previously smelling the odorant. The degree of affect experienced is the arithmetic mean of decisions to open the gate in either the positive or negative direction, weighted by the degree to which each object encoding in the store is activated. Thus the past consequences of experience with an odor dictate the hedonic response, but this is lagged against dramatic changes by the stochastic nature of the odor activation process. That is, not all object encodings will be activated (and varying by degree too) on each encounter with a particular odor, preserving some gate decisions (positive vs. negative) from interference. Unlike the intensity and quality models presented above, where there is some precedence for the modules design, we have very little to go on with olfactory hedonics. There are only very limited neuropsychological data, which does not inform us as to the correctness of this model (but see below). Although it appears that evaluative conditioning with odors is sensitive to changes in hedonic consequences, which this model can accommodate, the actual malleability of odor hedonics is still largely unstudied. The three basic aspects of this model of olfactory information processing are represented in figure 7.4.

Figure 7.4 represents a model unelaborated by the findings described in this chapter. Needless to say even this unelaborated model can not claim to be anything other than a crude representation of how the olfactory system extracts meaning from the chemical array. It does, however, have the power to both explain a considerable number of findings and to make predictions as well. In brief, it places a premium on experience and it suggests a dissociation between intensive perception and qualitative perception, which has at least been observed. In terms of prediction, it suggests, first, that hedonic responding must depend on the presence of the object store and, second, that it should be possible to find brain-injured patients who demonstrate intact quality and intensity perception but with abnormal affective responses to odors. As we noted, the neuropsychology of olfaction has little as yet to say about olfactory hedonics. Likewise, a really systematic search for the pattern

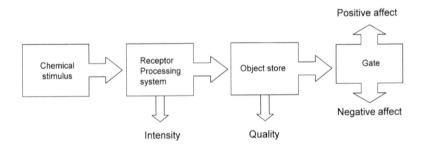

Fig. 7.4. Basic aspects of human olfactory information processing.

of dissociations suggested by this model, notably between quality and intensity, has not yet been conducted.

This now brings us to the point where we can examine how the model presented in figure 7.4 might be elaborated to deal with what we know about human olfactory information processing. The obvious place to start is with attributes most directly connected with odor memory. Clearly some capacity to briefly store olfactory information must be present. Assuming that we do not rely solely on semantic mediation, and we clearly do not, how else could we solve problems like those posed by the triangle test that involve serial comparisons, each separated by a brief interval? The type of model of short-term memory proposed by Cowan (1988) is one obvious contender. Effectively it allows us to use the extant model to explain the brief retention of information, through the process of activation and decay. That is when an encoding is activated, the period of activation lasts longer than the initial period of stimulation, resulting in the ability to retain information longer than the period for which the stimulus is directly available. Note, this residual activation is outside of conscious awareness. Such a system is affected by capacity, *but* only in so much as a greater number of odorants will mean a greater likelihood of similarity between members of the set. As similar odorants activate similar and overlapping sets of encodings, this will result in interference and thus degrade recognition accuracy.

Exactly the same explanation can be advanced for tasks involving longer delays followed by recognition of previously smelled odorants. All we have to specify is that activation decay may take a long time—a very long time indeed. Consequently, an odorant may be recognized over short or long intervals by virtue of its existing state of residual activation. This ability, which might extend for weeks or months, is subject to exactly the same constraints as that imposed on briefer durations. It is beneath the level of conscious awareness and increased set size produces interference, by virtue of greater likelihood of similarity between set members. Not surprisingly, we would predict that exactly the same factors that influence short-term tasks should equally influence long-term tasks, namely, that both are capacity limited by virtue of similarity constraints. The available evidence is consistent with this, although to our knowledge nobody has as yet directly contrasted this at short and long delays within the same experiment.

A significant feature of olfactory information processing is the ability to form associations between olfactory objects and other events, be they words, colors, or autobiographical memories. To achieve this, connections would need to be formed between the activated odor object encodings and other discrete memory storage systems (e.g., semantic memory). Such associations appear to form in at least two ways. First, by effortful association, typical of experiments in which, for example, participants learn to associate an odor with a name (e.g., Stevenson and Dempsey 2002) or place (Takahashi 2003). Second, by incidental association, such that an odor present at study and again present at test, acts to facilitate retrieval of information that was also present during the study phase (Schab 1990). Both of these processes probably call upon related mechanisms but result in somewhat different outcomes. The first obviously requires practice, in that the odor and the to-be-associated cue need to co-occur several times for participants to be able to perform effectively on demand. This may be a slow process because, as we already noted, even a familiar odor may activate several encodings, and all of these in a somewhat different manner and degree each time the odor is encountered. Thus there may be several associations formed of differing strength from each of the activated encodings to the cue. The second, incidental learning may reflect the same process, namely, that associations form between activated encodings and other information that is present currently, be it visual, auditory, perceptual, or semantic. However, because the associations formed are likely to be idiosyncratic—that is, only activated encodings will be associated and those to the degree to which they were activated—reactivation of all the associated memories will *also* be idiosyncratic, that is, a very close match of olfactory encoding activations will be needed to reinstate the links formed earlier. Consequently this type of information is likely to be highly preserved for long periods, because it is unlikely that *exactly* the same stimulus will be encountered again and, even if it is, it will not always result in the same pattern of activations. Thus such encodings are likely to be resistant to interference and long-lasting, assuming limited trace decay. Although this is an easy fit with the model it fails to account for at least two types of findings. First, it is directionally neutral, that is, associations should be as hard to form from an odor to a nonolfactory cue, as from a nonolfactory cue to an odor. But as we saw earlier this is not the case (Bower et al. 1994) and the model suggests no obvious reason why this should be so. Second, the associative interference data are problematic. On the one hand incidentally acquired odor-room associations are extinguishable (Koster 2002). Although the model does not appear to suggest this, this finding is not incompatible. Most odorants are not chemically identical, as they were in the Koster et al. (2002) study. Where they are identical, they are presumably more likely to activate the same encodings in the object store. Because the model does not specify that learning new associations is especially hard; rather, existing associations tend to be preserved by virtue of the probability of encountering similar encoding conditions, the Koster et al. (1992) results can be explained. This then raises the problem of the findings of resistance to interference by Lawless and Engen (1977) and Dempsey and Stevenson (2002). The difference here may be that in each of these cases the learning was not incidental; thus, the resulting associations were considerably more robust. Arguably then, the failure to find retroactive interference may have been a consequence of test sensitivity, that is if sufficient retraining had taken place (especially in Dempsey and Stevenson 2002), then interference might eventually have been obtained.

One of the most intriguing aspects of olfactory information processing is why odor naming is so hard. In one sense we have already offered an explanation for this, because most name learning must presumably rely on incidental learning. Consequently, the process will be idiosyncratic as described above and on many occasions different patterns of activation will occur with a similar odorant, which are different from those made when its name was available. It then becomes a problem of encoding specificity, namely, a mismatch between retrieval and encoding conditions. However, if odor names are learnt effortfully, this situation should markedly improve by virtue of increasing the probability that the odors pattern of activation will match those that were present at the time of encoding the name. This at least is consistent with the findings from several studies (e.g., Cain 1979). One prediction from this account, which is born out in the literature, concerns the tip-of-the-nose phenomenon (Lawless and Engen 1977). When participants are unable to name an odor, in general, they are able to report the broad class from which the odor is drawn (e.g., fruity). This would be expected if there were general connections to the area of semantic memory dealing with this category, but a failure to activate the precise link to the name residing within that area. That is, the tip-of-the-nose is a further consequence of the stochastic nature of the activation of odor objects.

A related issue concerns the apparent difficulty that most participants have in evoking odor images. As we noted previously, participants who engage in paired-associate learning between the odorant and cue, prior to attempting to form an image of the odor when its cue is presented, nearly always (at least in the published literature) demonstrate imagery-like effects (e.g., Algom and Cain 1991; Algom, Marks, and Cain 1993; Stevenson and Prescott 1997; Djordjevic, Zatorre, and Jones-Gotman 2004; Djordjevic, Zatorre, Petrides, and Jones-Gotman 2004). Arguably this depends on the same processes as before, that is, connections between an odor and a name are formed incidentally and are likely to result in only weak activation of an odor object by virtue of the fact that the associations are only to a subset of all the possible activations that might be expected to occur if the odorant was actually smelled. Consequently, deliberate strengthening of these associations results in activation of most of the encodings that would normally occur. Thus difficulty of evocation is a consequence of the same processes that result in poor odor naming.

To what extent does our model constrain an explanation for imagery? One concern that we have raised previously (Stevenson and Boakes 2003) is that the olfactory system may be unable to discriminate between top-down and bottom-up generated images, which could result in possibly fatal confusions—did I imagine that smell or is it real? We used this to argue against a capacity for imagery in a model very similar to the one presented here. Two things have changed since then. First, the evidence favoring percept-based imagery is much stronger than before. Second, there may be another means of reality monitoring that stops us from confusing a volitional image with a stimulus-generated representation and that is the absence of affect in odor images. Although this has not been directly investigated as yet, odor images do not in general involve spontaneous reports of hedonic attributes as real percepts and hallucinations do (Stevenson and Case 2005). If this observation turns out to be correct, then it is inconsistent with the model, because the model does not preclude the generation of affect from a top-down-only stimulus. One way round this is to suggest that the affect system has its own separate recognition system, something that is supported by the only study to investigate this (Perl et al. 1992; see chapter 5). A more definitive answer will come from examining whether patients who have no ability to experience odor quality (i.e., lack the object recognition module) can still experience appropriate hedonic responses. Unfortunately, this was not investigated in HM or in any of the related neuropsychological data cited earlier.

The final issue that we wish to examine is priming. Priming should, according to what we have discussed so far, be enabled by two methods. First, if an odor and a name have been well associated, then incidental presentation of an odor's name may result in priming of its olfactory encoding, that is, activation, but to a lesser extent than would be the case with odor imagery. Second, incidental exposure to an odorant should also result in priming, this time by virtue of the gradual decay in activation of the encodings produced by smelling the odorant. In both cases this should result in enhanced recognition of the odorant at some subsequent point in time. Notice how similar this is both to the process involved in recognition memory and in imagery, the difference being either the mode of activation or its degree.

In summary, the basic model presented in figure 7.4 actually requires remarkably little elaboration to cope with a broad range of findings. The two key changes are the rate at which activation may decay and the ability to form associations. Neither of these seem particularly radical, yet they gain the model considerable explanatory power. This elaborated model is also predictive. The model implies that priming should occur, it also suggests that interference in paired-associate learning procedures should be a direct consequence of the effort put into strengthening the association. In this respect, odor-cue associations should behave like other forms of successive paired-associate learning, albeit somewhat slower. The model also indicates that neuropsychological dissociations will not be obtained between short- and long-term olfactory memory tasks, and that such tasks should be subject to similar constraints if tested within the same experiment. Models of perceptual and cognitive function have been useful in other domains in shaping the research agenda, irrespective of their ultimate correctness. Hopefully, the model here should have a similar effect.

Conclusion

The elaborated model of olfactory information processing presented above attempts to unite findings reported in the earlier part of this chapter concerning short- and long-term recognition memory, associative learning, and imagery. This body of findings, accumulated mainly during the past thirty years, has to date been rather disunited. To our knowledge, apart from applying general models of memory, there has been no attempt to provide a cohesive framework that links olfactory perception and cognition in the way that we have here. Obviously, and this must be restated, the model is only a preliminary sketch and it relies on many assumptions that have yet to be tested. Perhaps therein lies its strength: it does make quite clear predictions that span from neuropsychology to learning.

Odor Memory in Nonhuman Animals

Given the central role of odors in the lives of many (most) nonhuman animals, odor memory has been the focus of extensive research. Odor memory in animals has been examined in both naturalistic paradigms (e.g., foraging behavior, kin recognition, homing) and in laboratory settings, including both implicit and explicit memory paradigms. Behaviorally, odor memory in rodents has proven to have striking similarities to higher-order cognitive performance in nonhuman primates and humans. In addition, the anatomy and physiology of the olfactory system have proven ideal for studies of the mechanisms of odor memory—relatively simple cortical circuits with well-described synaptic organization and synaptic pharmacology and strong ties to higher-order memory structures such as hippocampus and fontal cortex.

In this section, we will outline both behavioral models and neural mechanisms of odor memory in animals. The goal will be to place the implicit, perceptual olfactory learning described in previous chapters in the context of a broader view of odor information processing and explicit memory. A large literature exists on the role of the rodent hippocampus in explicit memory for odors. This section will primarily focus on rodent work, though several striking analogies with invertebrate work will be noted.

Explicit Odor Memory

Although the first half of this chapter paid considerable attention to odor naming in humans, a direct comparison with this phenomenon cannot be made in animal work. Animals can and do easily ascribe associative meaning to odors, though linking a symbolic label to odors (naming) is unlikely. The ability to assign meaning or predictive value to odors through associative (either operant or Pavlovian) conditioning has been demonstrated in a variety of invertebrate (honeybees, fruit flies, terrestrial mollusks, moths, spiny lobster) and vertebrate (salamanders, rodents, birds, fish, primates) species. Most species examined to date can be easily conditioned to prefer or avoid odors associated with biologically significant events.

Simple "Go, No-Go" odor discrimination learning is readily demonstrable in rodents after relatively few conditioning trials. Discrimination learning is robust under a variety of conditioning paradigms and reward schedules. Learning a simple discrimination generally occurs quickly, is resistant to both proactive and retroactive interference, and is stable for weeks or months. Thus, for example, learning A+ versus B− (where responding to odor A+ with some appropriate operant such as nose poke, bar press, or maze arm selection results in reward, whereas the same response to B− results in no reward) can occur within twenty or fewer trials. Once acquired, memory for this odor pair can last many weeks (long-term memory), even if in the intervening period scores of other odor pair discriminations are learned (i.e., minimal retroactive interference). Similarly, having learned multiple pairs of odor discriminations does not impair subsequent learning of new pairs (i.e., minimal proactive interference). Staubli and colleagues suggest these characteristics of olfactory data memory are similar to learning lists of words or faces in the auditory and visual systems of humans and also appear similar to human olfactory memory described above.

Two-odor discrimination tasks most commonly involve successive cue sampling, that is, one odorant is presented at a time and a decision is made by the animal as whether that odor is the rewarded, S+ odor and thus should be responded to (Go), or is the unrewarded, S− odor and a response should be withheld (No-Go). This task appears to be very simple for normal rodents, although it is probably not representative of real-world situations. Another form of the two-odor discrimination task is a simultaneous cue-sampling task, where both the S+ and S− odors are presented and can be sampled within the same trial and the subject must choose the S+ odor from the two. This task may more accurately reflect the common experience of an animal faced with multiple stimuli simultaneously and respond appropriately. Rats learn this task rapidly, though perhaps not as well as the successive cue task. As discussed below, these two-odor discrimination tasks appear to involve different internal representations and thus rely on different neural circuits for their expression.

In two-odor discrimination tasks, rats learn and remember the associations of both the S+ and the S− odors. Using a previously learned S− odor as the S+ in a subsequent odor discrimination tasks (reversal learning) produces initially very poor performance (below chance) as the animals must unlearn/extinguish their previous associations with the previously S−, now S+ odor.

S+ Odor	S− Odor	Trials to criterion
A	B	100
C	D	80
E	F	30
G	H	1
I	J	1

Fig. 7.5. Animals improve in performance on a series of simple odor discrimination learning tasks, eventually performing near perfect on new odor pairings. The animals develop a win-stay, lose-shift strategy, demonstrating a learning set or learning-to-learn ability.

Odor discrimination performance can be dramatically improved by previous learning about different odor pairs. This improvement in the ability to make simple discriminations over repeated new odor pairs is called set learning or learning to learn and, in general, is regarded as a higher-order cognitive function. The animal learns that one odor of a pair will always be rewarded while the other is not and thus learns a win-stay, lose-shift strategy, allowing one trial learning on later odor pairs. Specific brain regions appear to be differentially involved in odor set learning distinct from basic odor discrimination learning, as discussed below. Similar learning sets do not appear to be acquired by using other sensory systems in rodents, and thus it has been argued that this reflects the unique access of the rodent olfactory system to higher-order cognitive systems such as frontal cortex (via the dorsomedial nucleus of the thalamus and piriform cortex).

Another characteristic of simple odor discrimination learning is that odors are learned as single stimuli even if composed of multiple components, as demonstrated by Staubli and colleagues in a series of studies. As discussed in a previous chapter, rats were trained to discriminate two complex odor mixtures varying by a single component, for example, ABC+ versus ABD−. The task could have been solved by elemental analysis of the odor mixtures through focusing on C+ and D− and ignoring the overlapping components AB. Alternatively, the task could have been solved by developing a synthetic, gestalt representation of ABC and ABD, distinct from the components. The latter strategy seemed to occur because there was no transfer of learning ABC+ versus ABD− to performance in discriminating C+ versus D−. The latter task appeared to be treated by the rats as a new, novel discrimination problem. These results fit with our view of odor object perception as outlined earlier.

Note, however, that how binary odor mixtures are perceived is strongly influenced by the nature of the components and their relative concentrations. Synthetic or gestalt perception of binary odor mixtures can be impaired if the odorant components are highly dissimilar in either molecular conformation or relative concentration, and depending on the task requirements, prior experience and expectations. For example, a strong component can dominate the perception of the binary mixture and, in turn, be the focus of associative conditioning, blocking association with the more minor component. Similar effects can be observed in multimodal mixtures that include odors. If rats are trained to a multimodal mixture of a visual stimulus and an odor, the odor component of the mixture dominates, reducing and overwhelming conditioning to the visual stimulus (c.f., odor-visual cue paired-associate learning in humans discussed earlier).

Furthermore, the ability of an odor to function as a conditioned stimulus in an associative conditioning paradigm is influenced by innate odor meaning, essentially a biological constraint on olfactory conditioning. Thus, for example, moths (and other invertebrates) can be conditioned to approach novel odors paired with reward or express other specific conditioned behaviors such as proboscis extension. However, male *Spodoptera littoralis* were unable to learn a proboscis extension response to the female sexual pheromone. In contrast, female *S. littoralis*, which do not express an innate response to the female pheromone, could be conditioned to this odor. This suggests that the innate male reflexive responses to female pheromones limit new associative meanings for this odor.

A final point about simple odor conditioning relates to the apparent unique relationship between odors and tastes. Novel odors can be conditioned to acquire either an appetitive or aversive association. However, aversive odor conditioning can be potentiated if the novel odor is paired with a novel taste during aversive conditioning. Thus, pairing a novel cherry odor with LiCl injections that produce illness will result in a subsequent aversion to cherry odor in rats. However, if the novel cherry odor was experienced in the presence of a novel sucrose taste prior to the LiCl injection, the subsequent aversion to cherry will be much stronger. The taste potentiation of odor aversions implies a special relationship between odors and tastes. As described in more detail in other chapters, this relationship can come to be expressed perceptually as the learned experience of sweet odors.

In addition to simple odor discrimination and set learning, odors can be used for delayed-match-to-nonsample (DMNS) and paired-associate learning. The delayed-match-to-sample task involves presenting an odor (the sample) and then after some time delay (usually corresponding to short-term memory traces described in humans above) asking the animal whether a new stimulus matches or does not match the original sample. After an additional delay, a new sample can be presented and the task repeated. Thus, this is a classic working memory task where ideal performance involves remembering the sample for the duration of a single trial, but then clearing memory of that sample prior to the start of the next trial. Two types of olfactory DMS tasks are used in rodents. Continuous DMS uses the test stimulus of one trial as the sample stimulus of the next trial, whereas standard DMS uses two new stimuli in each successive trial.

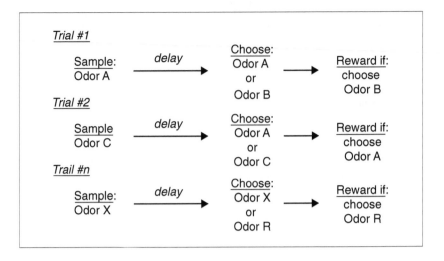

Trial #1

Sample:
Odor A → *delay* → Choose:
Odor A
or
Odor B → Reward if:
choose
Odor B

Trial #2

Sample
Odor C → *delay* → Choose:
Odor A
or
Odor C → Reward if:
choose
Odor A

Trail #n

Sample:
Odor X → *delay* → Choose:
Odor X
or
Odor R → Reward if:
choose
Odor R

Fig. 7.6. Delayed-match-to-nonsample (DMNS) tasks are commonly used to test short-term or working memory for odors. The animal is presented with an odor sample, which is then removed for a specified delay of seconds to minutes, followed by a choice test between the sample and a new odor. In delayed-match to nonsample, the animal must choose the new odor (nonsample) to receive a reward. After another brief delay, a new sample is presented which may or may not be related to the stimuli in the previous trial. The animal must disregard contingencies in previous trials to perform accurately in the current trial.

In a continuous DMNS task rats performed well above chance with delays of at least 60 seconds, roughly comparable to performance of monkeys on a DMNS visual task. Furthermore, performance was influenced by the size of the odor set from which samples and test stimuli were chosen (c.f., human memory performance described above). Very small odor sets (2 odors) degraded performance in this paradigm, whereas performance was much better with larger (e.g., 8 or 16) odor sets. Rats also perform well in a standard DMS task.

In a modification of the DMNS task, Eichenbaum and colleagues examined memory span for lists of odors. Dudchenko et al. (2000) presented an odor sample in a cup placed in an arena for the rat to sample. The rat was removed for a delay period, then returned to the arena where now a second cup scented with a new odor was present and the rat had to choose the cup with the new (nonmatch odor) to receive a reward. The rat was again removed from the arena, a third cup scented with a third odor was added, and the rat had to choose the new odor from among the three. This was repeated up to

25 times, requiring the animal to remember up to 25 sample odors to determine which was the new, nonmatch. Performance on this task decreased only slightly over the 25 items, indicating a memory span of at least 25 odors in this task.

In addition to having a large memory span for odors, rats can also learn and remember odor sequences. Memory for odor sequences has been examined in at least two distinct paradigms. Agster, Fortin, and Eichenbaum (2002) trained rats to respond to odors in a particular sequence, such as A-B-X-Y-E-F, where each odor (represented by a letter) was presented in a two-odor choice task. If odor A had been the correct choice on the previous trial, then odor B would be correct on the current trial, and X would be the correct choice on the next trial. Rats learn this task and can even perform well on a more complex task in which choices are made between two overlapping sequences. For example, rats learn two sequences—odors A-B-X-Y-E-F and odors L-M-X-Y-P-Q. As can be seen, the two odors in the middle of the sequences are the same. Knowing which odor to choose on the trial after Y was the correct choice depended on what came at the beginning of the sequence (if A-B, then choose E after Y, if L-M then choose P after Y). Given several

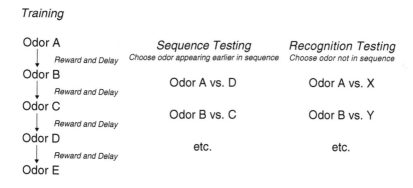

Fig. 7.7. Memory for odor sequences can be examined using a task developed by Agster, Fortin, and Eichenbaum (2002). Animals are trained with a temporal sequence of odor-reward pairings, where, for example, odor A always precedes odor B (delays between odors are a few minutes). After training, animals can be given choice tests to determine whether they remember the sequence of odors by rewarding choices of odors that occurred relatively earlier in the sequence (e.g., choose odor A when given a choice between A and D). Animals can also be tested for basic recognition of odors in the sequence by rewarding choices for novel odors not present in the original sequence (e.g., choose odor X when given a choice between A and X).

months of training, rats demonstrated over 80% correct performance. These data suggest that animals develop a representation not only of the current odor and its associations, but they can also develop memory for temporal patterns of odors, an ability that may be critical for olfactory-guided navigation.

Finally, odors can be used in a paired-associate task—a task classically used to examine declarative, explicit memory. Rats were trained to sample two odors presented in quick succession (0.75-second duration for each odor separated by a 0.5-second delay) and then required to respond if the two-odor samples were a trained pair and not to respond if the odors did not belong together. Thus, for example, if odors A and B were paired a response was rewarded and if odors C and D were paired a response was rewarded. However, a response when A was paired with C or when A paired with X was unrewarded. The temporal odor within the pair (e.g., A-B or B-A) was unimportant. Rats reached criterion performance on this task within 1,000 training trials. Similar paired-associate learning can be expressed by rats for odor-visual/tactile object pairs.

Together, this literature demonstrates that odors can be used by mammalian olfactory systems in very flexible, complex ways and that even invertebrate olfactory systems are capable of olfactory associative memory. Odors can acquire meaning through associative conditioning and can come to be associated with other odors or nonolfactory stimuli in stimulus-stimulus associations. Furthermore, information about the learned odors can be held and compared with other odors concurrently or previously presented. In other words, odors, composed of multiple, submolecular features, are treated as unique perceptual objects in declarative memory tasks. Depending on the complexity of the odor stimulus, features of the odor object can vary in saliency, be individually recognized, and/or gain independent associative value, but, in general, an odor mixture and its submolecular components are treated configurally as single objects. The situations promoting analytical processing of odors are not distinctly different from those operating in other sensory systems involving, for example, conditioning to or memory for visual objects.

Olfactory Priming and Odor Imagery

As described earlier for humans, exploring olfactory priming and odor imagery in animals is operationally difficult. In fact, in this regard the human literature has advanced far beyond that of the animal data in the area.

Nonetheless, there are some tantalizing electrophysiological findings that suggest an important role for experience-based expectation and top-down processing in rodent olfaction. In the mammalian olfactory bulb, activity within local circuits can produce oscillatory activity recorded in field potentials. These local field potential oscillations characteristically occur in a low-frequency band (theta, 4–10 Hz), an intermediate-frequency band (beta, 20–40 Hz), and a high-frequency band (gamma, 40–100 Hz). The gamma frequency oscillations are hypothesized to stem from mitral-granule cell reciprocal synaptic currents and demonstrate spatial patterns of amplitude across the olfactory bulb surface. One interpretation of these spatial patterns of olfactory bulb activity is that they reflect the odor-specific spatial patterns of olfactory receptor neuron input to the olfactory bulb glomerular layer. However, as Freeman and colleagues have demonstrated, these patterns can be changed with experience such as discriminative conditioning. The spatial patterns of mitral-granule cell activity thus reflect past associations and expectations, rather than simple odor physicochemical identity. In fact, in rats well trained to perform in an odor discrimination task, olfactory bulb gamma oscillations can begin several respiratory cycles prior to odor sampling, suggesting again an expectation and top-down component to this processing.

Neural Substrates

As noted above, olfactory memory, and memory in general, can be divided into several subclasses . These include (1) implicit memory such as habituation, sensitization, classical conditioning, and perceptual learning; (2) explicit memory such as delayed-match-to-sample paradigms, spatial memory, and paired-associate memory; and finally (3) set learning or learning to learn. Although specific behavioral paradigms are used to test each of these forms of memory, and evidence exists for the specific neural mechanisms underlying them (as outlined below), it must be emphasized that in a given context or situation several of these forms of memory may be evoked at the same time. For example, training in a delayed-match-to-sample task involves an explicit memory component to allow comparison of sample and test odor but, through exposure and familiarization with the stimuli, may also involve an implicit, perceptual learning component, modifying perception of the learned stimuli. Note also that the types of experiences involved may produce changes in a diverse collection of brain circuits (cortical and subcortical) that may ultimately be required for expression of the acquired memory. Here, we will emphasize neural plasticity within the primary olfactory system itself, though mention is made of circuits and mechanisms in other brain regions.

Implicit Memory

Implicit memory includes habituation, sensitization, perceptual learning, and classical conditioning. Habituation is a decrease in responsiveness to repeated or prolonged stimulation, relatively specific to the repeated stimulus, and subject to dishabituation. Habituation and adaptation allow sensory systems to filter background or currently nonsignificant stimuli, while maintaining responsiveness to novel stimuli. Habituation to odors could involve olfactory receptor adaptation and/or central mechanisms. In many thalamo-cortical sensory systems, cortical neurons adapt more rapidly and completely than more peripheral neurons. This could allow for rapid dishabituation if the contingencies change and the stimulus becomes potentially important. Similarly, although in the olfactory system both receptor neurons and mitral cells adapt to odors, piriform cortical neurons adapt much more rapidly and completely following either prolonged or repeated odor stimulation. Cortical adaptation is also more selective (displays less cross-adaptation) to familiar odors than mitral cell adaptation.

Anterior piriform cortical neurons rapidly adapt to novel odors, despite relatively maintained input from their excitatory afferent, glutamatergic mitral cells. A similar, rapid decrease in odor-evoked spiking occurs in the insect mushroom body neurons. Recent work has demonstrated that in the rat, this cortical adaptation may be due primarily to activation of presynaptic metabotropic glutamate receptors (mGluR) on mitral cell axons. The synaptic depression has a duration of less than 2 minutes following a 50-second odor exposure, as does the reduction of cortical responses to subsequent stimulation with that odor. Blockade of mGluR II/III receptors prevents afferent synaptic depression and cortical adaptation to odors. This presynaptic, homosynaptic mechanism of cortical adaptation allows maintained input to other, nonexperienced afferent inputs, and thus could serve as an ideal mechanism for gating continuous background input while allowing new or changing inputs to pass. (Other mechanisms may exist for state-dependent gating [Murakami et al. 2005] or long-term adaptation [Dalton 2000]).

Thus, during prolonged odor exposure, for example, a background room odor, olfactory receptor neurons begin to slowly adapt through a defined Ca^{2+}-dependent mechanism (Zufall and Leinders-Zufall 2000), mitral cells begin to slowly adapt through an unknown mechanism, and at the same time, piriform cortical neurons rapidly adapt, in part, through a mGluRII/III-mediated decrease in glutamate release from mitral cell axons (Best and Wilson 2004). The more peripheral and receptor changes most likely function to maintain receptor function within an optimal dynamic range rather than serving to filter background odor. With cortical adaptation, if the odor becomes significant or arousal level increases (dishabituation), central olfactory system responding may return (Scott 1977). One mediator of this dishabituation could be norepinephrine from the locus coeruleus. Norepinephrine has at least two actions related to adaptation in the olfactory system. First, norepinephrine can increase mitral cell responsiveness to olfactory nerve input (Jiang et al. 1996). Thus, in rats, a tail pinch can reinstate mitral cell responses to an adapted stimulus (Scott 1977), perhaps in part via a noradrenergic enhancement of mitral cell responsiveness to olfactory nerve input. In concert with this change in the olfactory bulb, norepinephrine can modify/enhance piriform cortical neuron responses to odors (Bouret and Sara 2002). In fact, norepinephrine can block afferent synaptic depression to cortical neurons, perhaps via a direct interaction between noradrenergic β-receptor activation and mGluRII/III receptors (Best and Wilson 2004). Acetylcholine (ACh) has also been implicated as modulating odor habituation, although precise mechanisms have not been determined.

While this mechanism can account for the rapid changes in cortical and behavioral odor sensitivity that occur with exposure (Best et al. 2005), it does not account for long-term changes in odor sensitivity that occur after more prolonged exposure. Long-term (weeks) odor exposure can produce both decreases (adaptation) and increases (sensitization) in behavioral odor responsiveness (Dalton and Wysocki 1996). The mechanisms described above last only minutes, whereas long-term exposure effects can last months or years. The persistent effects of long-term odor exposure may reflect changes at the receptor sheet (Wang, Wysocki, and Gold 1993), differential neuronal survival in the olfactory bulb (Woo, Coopersmith, and Leon 1987; Rochefort et al. 2002) and/or changes in cortical circuitry.

Another form of implicit memory is perceptual learning (see earlier chapters). Prior experience and/or familiarity can enhance sensory acuity for familiar stimuli compared with sensory acuity for novel stimuli. Perceptual learning has been described in most sensory systems in addition to olfaction. Several mechanisms may contribute to olfactory perceptual learning. For example, simple odor exposure modifies olfactory bulb circuit and mitral cell single-unit responses to subsequent odor presentation at both a very rapid (millisecond) and at a long (hours to days) timescale. The rapid modification can occur within a single brief odor presentation and may reflect a dynamic fine tuning of odor-responsive ensembles (Laurent et al. 2001; Spors and Grinvald 2002). However, a long-term cascade is also generated that ultimately produces more permanent changes in subsequent mitral cell odor responses. Of most relevance to perceptual learning, mitral cell odor-receptive fields can shift toward the familiar odor, at least over short distances and along odorant feature dimensions such as carbon chain length (Fletcher and Wilson 2003). It has been hypothesized that these experience-induced changes in mitral cell odorant receptive fields are due to mechanisms related to either changes in olfactory receptor synaptic efficacy (i.e. long-term potentiation/long-term depression [LTP/LTD]), changes in mitral cell–inhibitory interneuron synaptic efficacy and/or changes in efficacy in cortical feedback. In support of changes in inhibitory connectivity, simple odor exposure in invertebrates results in progressive increase in local field potential oscillatory power, a measure noted as being sensitive to local interneuron synaptic inhibition. Regardless of the mechanism of receptive field change, the results suggest that familiar odorant features are encoded differently than novel features, with perhaps an enhanced representation of familiar features.

Simple odor exposure (odor enrichment) also enhances survival of newborn granule cells in the adult mouse olfactory bulb. Animals with enrichment-induced increases in granule cell number also have enhanced odor memory (Rochefort et al. 2002). Neurogenesis of olfactory bulb granule cells occurs throughout life (Rosselli-Austin and Altman 1979), and survival has recently been shown to depend on odor experience, with odor deprivation reducing survival (Najbauer and Leon 1995) and odor learning or exposure enhancing survival (Rochefort et al. 2002). It has been hypothesized that odor experience could selectively promote survival of newly generated granule cells into odorant-selective circuits, enhancing memory and discriminability for those odorants. Odor exposure during early development may also regulate survival of juxtaglomerular neurons near odor-specific glomeruli (Woo and Leon 1991), although this seems to only occur during early development. These data point to the potentially critical role of granule cell inhibitory interneurons in odor memory. In mammals, granule cells not only serve as lateral and feedback inhibition between output neurons, but they also are the primary recipient of centrifugal inputs to the bulb. There is a massive glutamatergic input from the olfactory cortex to the bulb that is very poorly understood. This pathway appears to be capable of experience-dependent plasticity (Stripling and Patneau 1999). In thalamocortical sensory systems, these descending inputs from the cortex back to the thalamus are critically involved in contextual and experience-dependent modifications of sensory coding. Cortical control of olfactory bulb activity looks like a rich area for future research (see chapter 3).

In addition to changes in the olfactory bulb, perceptual learning is also associated with changes within the piriform cortex. For example, piriform cortical neurons show similar levels of cross-adaptation to novel odors as mitral cells. However, after at least 50 seconds of familiarization to a previously novel odor, the ability of cortical neurons to discriminate between molecularly similar odorants is greatly enhanced. Both behavioral perceptual learning and this change in cortical discrimination can be blocked by the cholinergic muscarinic receptor antagonist scopolamine (Wilson, Fletcher, and Sullivan 2004). Muscarinic receptor activation enhances LTP of association fiber synapses (Hasselmo and Barkai 1995), thus scopolamine should reduce cortical synaptic plasticity. Based on theoretical and computational modeling work, it has been hypothesized that activity-dependent enhancement of intracortical association fiber synapses allows patterns of odorant features to be synthesized by cortical ensembles, creating perceptual odor objects from the collection of odorant features extracted by the periphery (Haberly 2001). Once these odor objects have been synthesized, discriminating objects from each other (rather than collections of overlapping features) is enhanced, accounting for enhanced acuity for familiar odors. Acetylcholine has also been implicated in the effects of experience on odor coding in the olfactory bulb (Linster and Cleland 2002) to enhance distinctiveness of evoked feature patterns.

The mechanisms described above can occur following simple odor exposure and familiarization. Associative conditioning through temporally pairing an odor with an unconditioned stimulus or reward also modifies the olfactory system. Classical associative conditioning, where an odorant signals the occurrence of, or is temporally paired with, an unconditioned or biologically significant stimulus has been examined in several paradigms. These include the associative conditioning underlying neonate recognition of maternal odors in rats, maternal recognition of neonate odors in sheep, learned recognition of mates in mice, and simple aversive conditioning. The results of all of these paradigms implicate similar associated neural mechanisms. First, the olfactory bulb (or accessory olfactory bulb) is modified by the conditioning, in general, with an enhancement of synaptic inhibition relative to excitation. Thus, for example, there is an enhancement of the ratio of γ-aminobutyric acid (GABA) release to glutamate release in the accessory olfactory bulbs of recently mated mice in response to the learned odor (Brennan and Keverne 1997). Furthermore, there is an increase in suppressive responses to the learned odor in mitral cells near the odor-specific activated glomeruli and a relative decrease in excitatory responses (Wilson and Sullivan 1994). Finally, there is enhancement (or at least modification) of local-field potential (LFP) oscillations, which are influenced by synaptic inhibition (Grajski and Freeman 1989; Martin et al. 2004).

In vertebrates, olfactory receptor neurons, mitral cells, and piriform cortical pyramidal neurons are all glutamatergic and most postsynaptic targets of these neurons express both N-methyl-D-aspartate (NMDA) and non-NMDA receptors. Recent in vitro work suggests that olfactory receptor neuron-to-mitral cell synapses can express LTP following high-frequency activation (Ennis et al. 1998). The precise role of LTP at this synapse in odor coding or memory is unclear, but it could shift the balance of excitation and inhibition within a glomerulus to enhance mitral cell output from previously activated glomeruli. Past experience can shift mitral cell odor feature tuning toward the familiar odorants (Fletcher and Wilson 2003), which could contribute to perceptual learning and enhanced odor acuity.

The mitral cell–granule cell synapse has also been strongly implicated as supporting long-term plasticity. Granule cells express both NMDA and non-NMDA receptors, and express high levels of calcium/calmodulin-dependent protein kinase II (CAMKII) (Zou et al. 2002) characteristics that are common in neurons capable of experience-dependent plasticity like LTP. Following associative conditioning there is a change in the ratio of glutamate release (presumably from mitral cells) to GABA release (presumably from granule cells) in favor of heightened GABA release (Brennan and Keverne 1997). This has been interpreted as an increase in mitral cell to granule cell synaptic efficacy induced by associative conditioning. In young rats, associative odor conditioning enhances mitral cell–suppressive responses to the learned odor (Wilson and Sullivan 1994), which has also been hypothesized to result from enhanced mitral to granule cell synaptic transmission.

Experience-induced changes in the balance of synaptic excitation and inhibition within the olfactory bulb circuit can not only influence the probability or rate of mitral cell action potential output, but can also influence spike timing and firing synchrony between ensembles of mitral cells. This is evident from pharmacological manipulations in invertebrates (MacLeod and Laurent 1996) as well as in work from transgenic mice (Nusser et al. 2001). Thus, an enhancement in granule cell GABAergic feedback to mitral cells in transgenic mice (which lack GABAergic inhibition of granule cells) enhances circuit synchrony as evidenced by increased oscillatory power in the olfactory bulb LFP (Nusser et al. 2001). Increased synchrony of olfactory bulb output neurons could further enhance the probability of LTP-like changes in synaptic strength within the olfactory bulb and/or within olfactory bulb efferent structures such as the piriform cortex. In accord with this interpretation, changes in olfactory bulb LFP oscillatory power in the beta (15–35 Hz) and gamma (35–90 Hz) frequency ranges are strongly correlated with odor learning and response to familiar odors (Grajski and Freeman 1989; Ravel et al. 2003; Martin et al. 2004).

These experience-induced circuit changes, and the learned behaviors, depend on norepinephrine release in the olfactory bulb paired with the conditioned odor during learning (Gray et al. 1986; Sullivan and Wilson 1994). For example, infusion of norepinephrine or receptor agonists directly into the olfactory bulb is sufficient to induce changes in olfactory bulb odor evoked responses and behaviorally expressed memories of those odors. These results suggest that norepinephrine may convey information about the unconditioned stimulus to the olfactory bulb for convergence with odor-specific input. Both granule cells and mitral cells express noradrenergic receptors, and norepinephrine modulates the mitral-granule cell reciprocal synapse; however, recent work suggests a critical site of convergence between norepinephrine and odor input may be mitral cells. Activation of mitral cell β-noradrenergic receptors, combined with glutamatergic odor input elevates cAMP levels that ultimately result in phosphorylation of the cyclic AMP response element-binding protein (CREB) and subsequent protein synthesis changes that modify mitral cell function (Yuan et al. 2003). Hypothesized changes include mitral cell sensitivity to input and synaptic output that could account for the observed changes in olfactory bulb circuit function described above. Experimental modulation of cAMP levels or CREB phosphorylation produce the expected changes in learned behavior. Additional work will be needed to determine the effect of these manipulations on olfactory circuit function, but these results emphasize the critical role of centrifugal input to the olfactory bulb (and olfactory second-order neurons) in shaping olfactory processing and memory.

Very similar changes are observed in the insect antennal lobe following associative odor conditioning. Firing patterns and ensemble activity of antennal lobe second-order neurons are influenced by current context and past olfactory experience (Faber, Joerges, and Menzel 1999; Christensen et al. 2000; Laurent et al. 2001). Behavioral associative conditioning in honeybees requires the centrifugal neuromodulator octopamine within the antennal lobe

(Farooqui et al. 2003), similar to the requirement of norepinephrine for associative odor learning in mammals. In contrast to mammals, however, where spatial patterns of glomerular activation appear to be relatively stable following conditioning (Leon 1992), *Drosophila* and honeybee glomerular spatial activation patterns may be modified by associative conditioning (Sandoz, Galizia, and Menzel 2003; Yu, Ponomarev et al. 2004).

Most emphasis, however, in the study of neural mechanisms of associative odor conditioning in insects has focused on the mushroom bodies, a multisensory, higher-order central brain structure. The multisensory nature of inputs to the mushroom bodies allows for convergence of conditioned and unconditioned stimulus inputs. Lesions of the mushroom bodies disrupt associative memory in *Drososphila*. Molecular mechanisms of odor memory have been identified in the mushroom bodies of honeybees and *Drosophila* by using a variety of approaches (Dubnau and Tully 1998). In general, olfactory memory appears to activate and require an intracellular cascade in mushroom body neurons similar to that identified in mammals, including activity-dependent cAMP activation and CREB phosphorylation, presumably resulting in long-term synaptic and membrane plasticity.

As in insects, where higher-order olfactory structures are important for simple odor memory, increasing attention is being paid to the role of olfactory cortical, limbic, and neocortical circuits to simple odor memory in mammals. The hippocampus (Dudchenko, Wood, and Eichenbaum 2000), orbitofrontal cortex (Ramus and Eichenbaum 2000), and perirhinal cortex (Otto and Garruto 1997), which play critical roles in explicit odor memory described below, do not appear necessary for simple odor associative conditioning. In primates, however, single-orbitofrontal cortex neurons can respond to odors and tastes, and following conditioning involving odor discrimination for sucrose reward, a small subset of neurons appear to selectively encode the taste/reward association of the odor (Critchley and Rolls 1996). Similarly, in rodents, a small subset of orbitofrontal cortex neurons appears to encode the spatial locations associated with particular odors (Lipton, Alvarez, and Eichenbaum 1999). These results emphasize the multimodal nature of odor representations and the role of experience and learning in shaping those representations.

The amygdala, which receives a strong olfactory input from both the olfactory bulb and piriform cortex and is believed to be important in emotional memory and memory consolidation, is modified by odor-associative conditioning, as evidenced by learning-associated changes in evoked neural responses measured electrophysiologically (Rosenkranz and Grace 2002) and with c-*fos* immunohistochemistry (Tronel and Sara 2002). Activity within the anterior olfactory cortex is enhanced in response to learned odors (Hamrick, Wilson, and Sullivan 1993), though its specific role in odor memory in unknown.

However, changes in piriform cortex associated with simple odor-associative memory have been examined in the greatest detail. Single neurons in piriform cortex respond to a variety of apparently multimodal stimuli within an odor-learning task (Schoenbaum and Eichenbaum 1995; Zinyuk, Datiche, and Cattarelli 2001) and thus the opportunity for association of an odor with its context and associated consequences is present. Depending on the specific behavioral paradigm utilized, learning a simple odor discrimination can enhance c-*fos* labeling of anterior piriform cortex neurons (Datiche, Roullet, and Cattarelli 2001) and enhance both afferent synaptic strength and association fiber synaptic strength. Plasticity of association fiber changes may be longer lasting than those of the mitral cell input and may be most robust in the posterior piriform cortex compared with the anterior piriform cortex (Litaudon et al. 1997). In addition to changes in synaptic strength, membrane properties are modified in layer II piriform cortical neurons, resulting in enhanced excitability. However, these changes in membrane excitability may be more important in priming the cortex for learning subsequent odors, than in storing information about specific stimuli (Barkai and Saar 2001). Both the neural plasticity and the behavioral expression induced by training can be influenced by neuromodulators such as acetylcholine (Linster and Hasselmo 2001).

In both mammals and invertebrates, interactions between the bilateral olfactory pathways play an important role in simple odor memory. Both insects and adult rats can be unilaterally conditioned to odors such that odor stimulation of one nasal passage or one antenna can be paired with reward, and subsequently a conditioned response can be evoked by stimulation of either side (Kucharski and Hall 1987; Sandoz and Menzel 2001). Similar unilateral conditioning in neonatal rats, prior to the ontogeny of commissural fibers, results in unilateral access to the odor memory (Kucharski and Hall 1987). This bilateral access to odor memories suggests that central representations of simple odor memories are either distributed bilaterally or stored in structures that can be accessed by either hemisphere. In mammals, the piriform cortex is one such site, as single piriform neurons can respond to odors presented to either the ipsilateral or contralateral naris (Wilson 1997).

Bilateral access to odor memories would appear to be beneficial given that, at least in mammals where nasal patency fluctuates, an odor learned when one passage was closed could still be recalled by the other. Bilateral interactions may have additional consequences however, beyond simple access. For example, synthetic processing of odor mixtures in a learned discrimination task, and the blocking effect in honeybees both appear to require bilateral stimulation and processing (Komischke et al. 2003). Furthermore, training competing responses to the two olfactory pathways (e.g., A+ B− to the left nose and A− B+ to the right nose) in the terrestrial slug *Limax maximus* results in reduced total conditioning and active inhibition of olfactory system neurons (Teyke, Wang, and Gelperin 2000). Again, these results point to central bilateral convergent structures, beyond the olfactory bulb or antennal lobe, as being important for odor memories.

From the preceding descriptions it is apparent that learning about an odor changes even its very basic representation as early as second order neurons in the olfactory pathway. Simple associative odor conditioning in mammals relies on early parts of the pathway—the olfactory bulb and its direct targets such as anterior olfactory cortex, piriform cortex, and amygdala. Higher-order structures, which also receive strong olfactory input, such as orbitofrontal and perirhinal cortices and hippocampus, are not required for simple odor discrimination learning and implicit memory. Neuromodulatory inputs to the mammalian olfactory system, such as norepinephrine and acetylcholine, are required for simple odor memory, as are similar neuromodulators in insects. Furthermore, intracellular cascades (e.g., cAMP, CREB) involved in memory storage at the cellular level also appear highly conserved. The kinds of synaptic and circuit changes evoked by simple conditioning and implicit memory should influence not only spatial and rate codes for odor information, but also temporal patterning within ensemble networks, providing unique signals for familiar odors and those with learned significance. The reliance of simple odor memory on changes very early in the olfactory pathway suggests that the perceptual quality of odors may change with acquired meaning or familiarity. Learned odors just do not smell the same!

Odor Set Learning

As described earlier, learning new odors in an associative learning task facilitates the subsequent learning of other odors. Thus, for example, if a rat takes 50 trials to learn to discriminate odor A from odor B; subsequently, the same rat may only take 30 trials to learn odor C versus odor D, and eventually attain one trail learning on later odor pairs (Slotnick, Hanford, and Hodos

2000). This is referred to as set or rule learning and is robustly displayed by rats. Although many cognitive processes may participate in odor set learning, a more simple contributing mechanism could be an experience-induced change in plasticity or potential for plasticity within the same neural circuits believed to be involved in simple odor conditioning for the first odor (potentially a form of metaplasticity?).

Recent work in the piriform cortex has provided evidence for just such a mechanism for odor set learning. Barkai and colleagues have found that for several days after learning an odor discrimination task, piriform cortical neurons are modified such that they demonstrate increased excitability, decreased after-hyperpolarizations, and a potential electronic shortening of apical dendrites (Saar and Barkai 2003). This combination of biophysical changes results in cortical cells that may fire longer and more intensely to sensory input, making associative synaptic plasticity more likely, and thus facilitating learning of subsequent odor discriminations. A similar combination of effects can be induced in naïve cortical cells through activation of ACh muscarinic receptors, and ACh is important in the emergence of set learning behaviorally (Saar, Grossman, and Barkai 2001). These data suggest that initial learning primes the piriform cortex through a cholinergic mechanism into a sensitive state for heightened learning of new discriminations. The results thus also argue that the piriform cortex could play a critical role in the apparently higher-order memory function of set learning.

Explicit Memory

As described above, explicit memory includes memory for facts that can be cognitively manipulated: it is evidenced in behavioral paradigms such as delayed-match-to-nonsample and paired-associated memory. Much less is known about specific neural mechanisms of these types of olfactory memory, although brain regions involved have been identified. Neurons in the piriform cortex, orbitofrontal cortex, perirhinal cortex, ventral striatum, and hippocampus may contribute to memory and performance in olfactory explicit memory tasks (Slotnick 2001). Work with these paradigms emphasizes three main points: (1) explicit memory is distributed across both primary sensory and higher order regions; (2) even neurons in traditionally primary sensory areas, such as piriform cortex, may respond to multiple, nonolfactory aspects of the conditioning paradigm such as stimulus hedonics and nonodor task-related stimuli; and (3) it is likely that much of the implicit memory-associated changes in the olfactory system outlined above also occur in these paradigms.

Lesions of regions such as the hippocampus, orbitofrontal cortex, and perirhinal cortex disrupt performance in explicit odor memory tasks, while generally leaving performance in simple Go-No Go odor discrimination tasks and associative conditioning intact (e.g., Otto and Eichenbaum 1992). This suggests that these higher-order structures are required for explicit odor memory, as they are for explicit memory of other, nonolfactory information in rodents and primates. However, many of the same types of multimodal stimulus processing important for explicit memory and expressed by neurons in higher-order structures are also expressed within the primary olfactory system. Thus, Schoenbaum and Eichenbaum (1995) reported that neurons in both the rat orbitofrontal cortex and piriform cortex respond not only to odors, but also other aspects of a simplified paired-associate odor discrimination task such as pre-odor-sampling cues, odor-reward associations, consequences of the previous trial, etc. In fact, no major differences were obvious between the orbitofrontal and piriform cortical neurons with the techniques utilized. The differential inputs and projections of orbitofrontal cortex and piriform cortex may provide these two structures with different influences on learned behaviors, but there appear to be striking similarities in the kinds of processing these two very distinct structures express.

Explicit olfactory memory may, in fact, rely on changes not only within structures such as the hippocampus and orbitofrontal cortex (and primary sensory cortex), but also on changes in communication between these structures. Each component of this multisite circuit may contribute something slightly unique to the information to be learned, e.g., changes in piriform cortex may enhance the discriminability of the learned stimulus whereas changes in the orbitofrontal cortex and amygdala may encode multimodal associations and predicted consequences of the odor. Explicit, flexible memory of the odor then may require changes in communication between these different networks (essentially a more global ensemble representation). As in other systems, emergence and functioning of large-scale ensembles of this type could be facilitated by local circuit oscillations so apparent in olfactory and limbic circuits. Synaptic plasticity within these circuits could not only modify transmission efficacy between pairs of neurons, but also modify temporal dynamics of local and large-scale oscillations.

As noted, this view emphasizes that, although tasks can be developed that are largely indicative of explicit memory, these tasks most likely evoke both explicit and implicit memory consequences and mechanisms. Thus, while learning to expect that orange odor signals a sucrose reward, but only in a specific context and when following sucrose-rewarded cherry odor stimulation (an explicit memory task), the animal is also altering its perceptual representation of orange odor (implicit, perceptual learning). Damage to the hippocampus may impair the explicit memory, but it is also likely that a disruption of perceptual learning could affect performance in this task. The interplay between simple implicit memory and more complex explicit memory is an area needing further research in olfaction.

Summary of Findings from Animal Work

Both associative and nonassociative implicit memory are correlated with changes in olfactory bulb, piriform cortex, and limbic regions such as the amygdala. Thus, the neural substrates or consequences of even very simple memory are distributed in multiple brain circuits. Just as multiple circuits are involved in odor memory, so there are multiple synaptic and nonsynaptic mechanisms for neural change. These changes result not only in modified behavioral responses to the learned odor, but also in modified discriminability and perception of the learned odor.

Learned changes occur at least as early as second-order neurons of the sensory pathway (though long-term experience may change olfactory receptor expression also), and inhibitory interneurons play a critical role in expression of learned changes in circuit function. The important role of inhibitory interneurons to olfactory memory is similar to that of experience-induced changes in hippocampal and neocortical circuits, but is perhaps amplified in olfaction by the fact that olfactory bulb granule cells undergo neurogenesis throughout life and display experience-dependent survival.

Explicit memory and set learning abilities in rodent olfaction demonstrate the tight interrelationship between the primary olfactory system and the limbic system and higher-order neocortex. Rats express the ability to remember and manipulate olfactory information in much the same way that primates do with other sensory modalities. Given our understanding of peripheral odor processing and feature extraction by the olfactory receptor sheet, the explicit memory data strongly imply that odors are treated as individual, unique objects by the mammalian brain. Understanding odor perception and explicit odor memory thus requires some understanding of how complex molecular mixtures become perceptual objects.

The odor memory data from animal work also begin to emphasize the importance of bilateral interactions in olfactory function. Work in humans has suggested that there may be lateralization of some olfactory processing (Zatorre et al. 1992; Sobel et al. 1999). In the few studies in which it has been examined in animals such as rats and insects, intact bilateral interactions are required for some aspects of odor processing and memory. This may imply some lateralization of function where both sides, each performing slightly different tasks must communicate for normal function, or alternatively may reflect a type of mass action, where some minimal amount of processing power must be applied for normal function.

Finally, modulatory inputs, such as norepinephrine and ACh are critical for modulating both neural plasticity and behavioral change. This appears to be true for all three types of memory (implicit, explicit, and set learning), and although the specific neuromodulators differ, is true for both mammalian and insect systems. The role of neuromodulators in odor memory allows for regulation of information storage dependent on the significance of the olfactory experience, its context, and the internal state of the animal. Thus, a background odor may be less likely to be remembered than an odor associated with an arousing experience. Again, however, it is also the case that these neuromodulators not only affect plasticity, but also sensory physiology as early as the second-order neurons. Thus, mitral cell unit and olfactory bulb local field potential oscillation responses to food odors are enhanced in hungry rats compared to those of satiated rats, whereas responses to nonfood odors are unaffected by hunger, and this hunger modulation is mediated by neuromodulatory input to the olfactory bulb (Pager 1978; Chabaud et al. 2000). These state-dependent changes in firing rate and temporal structure of olfactory bulb output not only affect how the odors are encoded, but should also affect the probability of evoking synaptic plasticity in downstream sites and, thus, the chances that the odors and contemporaneous stimuli will more likely be remembered.

8

Implications

Thomas Kuhn (1962), in his *Structure of Scientific Revolutions* notes that facts are not like pebbles on the beach waiting to be picked up. Rather, he argues, how observations are interpreted (and thus become facts) is based on the theoretical viewpoint and resulting expectations of the viewer. Kuhn's argument has dual relevance here. First, we posit that smells (like facts) are not like molecules on the beach waiting to be inhaled, but rather are outcomes of highly synthetic, memory-dependent processing that is further modulated by expectation, context, and internal state. Second, how one interprets the experimental data in olfaction and, in fact, even the nature of the questions being asked, is strongly influenced by the theoretical view one has of the nature of olfaction. Data that do not fit conveniently into the existing theoretical view ("paradigm" in Kuhn's terminology) are either forced to at least superficially fit or commonly sidelined as anomalous.

For example, a physicochemical, analytical theoretical view of olfaction suggests a very clear line of investigation. Identify specific ligands for specific receptors, identify primary odor qualities, look for specific anosmias, attempt to induce specific anosmias through select receptor elimination or central lesions, and attempt to predict olfactory percepts from physicochemical structure (i.e., stimulus-response). We have reviewed historical (and more recent) efforts to address each of these questions, initial claims of success, and frequent subsequent contrary evidence (e.g., primary odors).

We suggest (humbly) that it is time for a paradigm shift in olfactory science, away from physicochemical views toward a synthetic, mnemonic view of olfaction as an object-based sense. The study of visual perception has had to undergo a similar process. As Richard Held (1989) notes, "during the nineteenth century it was apparently easy to believe that a correlation may be established between sensorineural and perceptual states. Knowing little about the neuronal processes left a great deal of latitude for speculation" (p. 140). However, Held goes on, "perhaps the stimulus provides only certain constraints on these processes. Perception is said to be distinguished from imagery by the presence of such stimulus constraints" (p. 142). Finally, Held concludes, "that conflict [between the physical nature of the stimulus, the initial sensorineural processing, and the ultimate percept] may be resolved by considering perception as reflective activity rather than passive reception" (p. 139).

In this final chapter we review, in brief, both the case for reflective, experience-based object recognition as the basis for understanding odor perception and the evidence for its operation via encoding novel olfactory experiences and matching inputs to them. We then argue that visual object recognition, which has been explored in considerable depth, offers some important parallels to the psychological and physiological processes that we have suggested underlie olfactory perception, most notably the reliance of all contemporary theories of visual object recognition on experience. The literature on visual object recognition is theoretically and empirically rich and in the penultimate part of this chapter we contend that a paradigm shift toward an object recognition view of olfaction can also provide similar benefits, by generating many testable hypotheses and encouraging theoretical development in this area. Finally, we end by briefly reiterating the central message that we would like to communicate—the perceptual ecology of smell suggests object recognition, the extant data support it, and what we need now is a concerted effort to find out if this perspective really is, as we think, the right one for understanding olfaction.

ODOR PERCEPTION AS MNEMONICALLY MEDIATED OBJECT RECOGNITION

The Ecological Perspective

Our first consideration must be the role that olfaction plays in the behavior of animals and humans and the implications that flow from this for function and process. In animals and humans, several common roles emerge, including recognizing food, mates, kin, predators, and disease vectors. The common element that underpins all these functions is the need to recognize the combination of chemicals that go to make up these various objects against a shifting and complex chemical background. Thus, the primary task the system faces is to learn what combination of features makes up a biologically significant pattern, encode this information, and then be able to discriminate this combination—the odor object—from both other objects and background sources of olfactory stimulation.

The same chemical stimulus may have radically different meaning depending on the receiver. In a limited number of cases this may be innate. The smell of cat odor to a rat induces a specific and hard-wired set of behaviors that are quite distinct from those observed for an aversive stimulus (say a trigeminal irritant) or for an odor that signals unpleasant consequences such as an electric shock (Dielenberg and McGregor 2001). On the other hand this same combination of chemicals—the cat odor object—may signal to another cat the presence of a potential mate or a close relative or that some other animal has infringed upon their territory. More typically, the meaning of an odor is dictated through experience, by associative learning between the odor object and what it signals. Conditioned taste aversions, the caloric value of food, disgusting fecal odors, the warning smell of methyl mercaptan (gas leak!), and the sweet smell of vanilla are all such examples. Learning and memory then play a pivotal role both in the response to many odors and in their passive acquisition as odor objects.

A basic strategy that appears to be deployed throughout the animal kingdom is the use of two overlapping modes of olfactory processing. The first mode is sensitive to specific chemicals or combinations of chemicals and generally produces either an innate physiological or behavioral response or a predisposition to respond in a certain manner. This labeled line or hard-wired mechanism probably reflects the oldest phylogenetic olfactory system, which can trace its descent back to the single chemical receptors present on the cell wall of unicellular organism such as *Escherichia coli* or *Paramecium* though is expressed in all animals. This mode of processing ensures rapid and appropriate responses to chemical signals that change little over time and that have major biological utility. Many pheromones would fall under this rubric, although the distinction is far from clear in many cases because of the often complex blends that go to make up certain pheromones and the fact that in higher animals, especially primates and humans, multiple sensory signals convey the same information as the olfactory pheromonal channel.

The second mode of processing, and the one with which we have been primarily concerned in this book, is also expressed in both invertebrates and vertebrates. This is the flexible, experience-based object recognition-processing mode. This mode is based on a large number of broadly tuned receptors, which allow the detection of a nearly endless combination of different odorants, by virtue of the relatively unique receptor output each chemical combination will generate. The key to this processing strategy is the ability to (1) encode these patterns of receptor output; (2) associate them with significant biological events; (3) subsequently be able to recognize and hence discriminate the same input pattern from other patterns and against varying chemical backgrounds; and (4) do so successfully even under conditions where the stimulus is heavily degraded. The basis for this type of processing is learning, that is, in acquiring both the olfactory input pattern, with subsequent pattern matching and forming associations with external events. The architecture underlying this processing mode must support degraded recognition (redintegration) and ensure the fidelity of its record of olfactory objects and associations. We briefly reiterate the evidence for this view below.

Evidence

In humans, a century of circumstantial evidence favoring an object recognition-processing mode has accrued, not because researchers were actively setting out to test it, but mainly because of the failure to progress with a stimulus-based model. Four lines of evidence are particularly telling. First, although the search for structure-quality relationships has clearly revealed associations between these two variables, *as there has to be*, it has consistently failed to explain why certain odors smell as they do. Relatedly, it cannot explain the large body of work that suggests that learning can significantly affect discriminability, intensity, and odor quality. Second, the systematic search for single chemical anosmias aimed to identify specific receptors tied to specific sensations, in the same way that anomalous color vision contributed to the identification of the three types of color receptor. We now know that most of these anosmias are partial rather than total, implying that it is rare for a single chemical stimulus to rely on one receptor type, and that the sheer number of specific anosmias identified suggests that a limited set of primary olfactory sensations is unlikely too. Third, studies of cross-adaptation reach a similar conclusion, as there is no clear support for the notion that odors that smell qualitatively similar cross-adapt, whereas those that smell qualitatively distinct do not. These types of findings imply that the perception of odor quality is more likely to be based on the overall pattern of stimulation than on the activation of a few relatively specific receptors. Fourth, linguistic evidence also argues against the existence of a limited set of primary odor sensations. The hierarchy of odor descriptors tends to be very flat, no consensus exists even among experts as to what qualities might constitute primary olfactory sensations, and systems of classification based on relatively few descriptors always fail to accommodate some odors.

Much, if not all, of the research findings described above took place in the pre–Buck and Axel era, that is, before we became fully aware of both the sheer diversity of olfactory receptors and their relatively broad sensitivity. Perhaps if all of this had been known earlier, then far more attention would have been focused on how the brain interprets the output from this system, rather than on the simpler (and perfectly logical) assumption that odor quality is wholly defined by physicochemical parameters. The sole focus on the stimulus as the key to unlock olfactory perception is clearly problematic. The stimulus may be encoded as a constellation of features, but the way in which this information is used by the brain—the focus of this book—is conceptually distinct from the labeled-line thinking that dominated much of the twentieth century's research into olfactory perception.

The mnemonically based object recognition process outlined in this book, not only addresses shortcomings of an analytical, stimulus-response model of olfaction, but also imposes some interesting features and constraints that should be evident in routine olfactory perception. One such feature would be the capacity to synthetically encode coincidental information (e.g., taste) that had a significant olfactory contribution (e.g., flavor). Not only is there physiological evidence from animals that multisensory encoding may take place in the putative store of olfactory objects, the piriform and orbitofrontal cortices, these structures also have the requisite connections to other brain areas to make this possible. Behaviorally, multisensory processing is seen to manifest in three ways. First, when an odor acquires taste-like qualities. Second, as with recent work in our laboratory that suggests that odors can also acquire trigeminal qualities, such as the coolness of menthol. Third, when a cue from another sensory modality primes or possibly even instantiates an olfactory percept.

One constraint in identifying a constellation of features as a distinct object is that access to the feature level description may be restricted. The input to the olfactory system is a pattern of glomerular activity (lots of features) from which an object (a constellation of features) is extracted. We suspect that the extraction process largely prevents access to the features; there may be three reasons for this. First, there may be little benefit in knowing what constitutes an object, because the object is the biologically significant entity—the level at which meaningful events are predicted. So there may have been little adaptive value in instantiating a more complex system that gives access to this information. Second, phenomenally, there is nothing to delineate a feature of the object from the object itself, so making the object the default ensures against misidentification. Third, if the features of the object are available for introspection, then one could argue that features from the chemical background should be available too. Yet, the very process of perception, in which a salient pattern is recognized, is an act of focal attention and the price that is paid for this may be the suppression of distracting information, such as features.

If a person or animal encounters a biologically significant odor and this co-occurs with some significant event, it is important that its meaning can still change (a good food later turns out to make you sick), but that the odor object—its phenomenal quality—is retained. This too appears to be the case. Perceptual memories of odors appear to be long lived in animals and humans, and in humans, they appear especially resistant to both retroactive interference and to extinction-like procedures. Conversely, there is some evidence from both humans and animals that the meaning attached to a particular odor is more malleable, in that it may be affected by counter-conditioning, but not by extinction-like procedures. Thus meaning may change but the encoding of the odor, required for perception, is more robust.

A further feature of an object-based recognition process is the ability to accurately identify an odorant under conditions of a degraded signal. Other perceptual systems are adept at this and it should be no surprise that a mnemonic-based object recognition system allows an animal or person to recognize an odor under conditions where the signal is incomplete. The most extensive evidence for this has been obtained from human olfaction, where presenting a component of a previously experienced odor mixture enables some perception of the whole to be experienced. This can be seen as analogous to the situation where an animal encounters a weak or partial signal for a conspecific, and that the fragment which is smelled is able to redintegrate the whole of the mnemonic encoding, so that what is experienced is the whole not the part. Clearly, this has significant biological utility.

So far we have focused on the object recognition process as being the key to understanding olfactory perception in contrast to a stimulus-driven approach to perception. It is worth reiterating here exactly how these two different approaches to the problem of olfactory perception interrelate, as we have frequently encountered the argument that perception *must* rely on the binding of chemicals to receptors, tacitly implying that any later processing is non-critical to a successful act of perception. There are several replies to this assertion. The first is that this assertion is correct but totally misleading. It is correct because this is obviously necessary to generate an input, but this input is modified considerably by the processes that we have described in this book, to the point where in some cases there is only a weak relationship between the stimulus and percept. The two are clearly correlated, but correlations do not imply causality. Second, by analogy, nobody would now accept that a study of light falling on the retina would provide a complete account of visual perception. Perception is a process that involves interpretation and the ascribing of meaning; it is not a stimulus-response system. Third, a stimulus-driven model has the major problem of explaining why there is any further processing of the signal beyond the receptor level. What point does the brain serve if the stimulus contains within it all the information needed to generate a response? Why, indeed, bother with a brain at all?

Comparing Visual and Olfactory Information Processing

All sensory systems, including vision and olfaction, share a common goal of extracting relevant information from the environment. To what extent do they share common information-processing features and where are the differences? As a starting point it is worth noting at least one crucially important difference. Visual objects are encountered in multiple orientations, all of which produce different two-dimensional patterns on the retina. In addition, they may be encountered under varying levels of illumination, partially occluded by other objects and at varying distance from the retina. Nonetheless, in most healthy participants, a familiar object can be readily recognized under a variety of different viewing conditions. Such view invariance has had a crucial impact on the types of theory that have been advanced to account for visual object recognition, and thus object perception, which we regard as synonymous. At least two general types of theory have been advanced to explain how the visual system achieves this. First, decomposition of the processed retinal output into primitive volumetric features and their interrelationships, with these then being automatically matched to memories of similar descriptions, result in object recognition (e.g., Marr and Nishihara 1978; Biederman 1987). Second, extraction of two-dimensional information about edges, surfaces, and vertices and then matching of this information to a limited repertoire of similar two-dimensional or three-dimensional feature-based descriptions again results in object recognition (Koenderink and van Doorn 1979; Ullman 1996). Both of these types of theory have received some experimental support and it is very likely that the visual system uses a combination of feature extraction, invariant relationships between features and multiple encodings obtained under different viewing conditions, for successful object recognition.

The similarities to and difference from olfactory object recognition are present at several different levels. First, the olfactory system does not have to deal with multiple orientations, levels of illumination, and occlusion, although it does have to deal with differences in concentration and with differences in the chemical context in which the target object is encountered. Although view invariance has been a powerful force in shaping theories of visual object recognition, a similar focus on odorant concentration has not occurred, in part, because there is still some dispute about the degree to which variation in concentration *can* affect odor quality within a constrained range and because, at least over larger ranges, there clearly are differences in perceptual quality. Thus, the genesis of theorizing in visual object recognition has been shaped to no small degree by the different demands that the environment places on the visual system, a useful warning of the importance of considering ecological factors in perception, especially the multicomponent nature of most stimuli.

Second, the role of experience in visual object recognition is significant in both classes of theory outlined above. For volumetric-based feature decomposition, although this is presumed to rely on identifying a set of three-dimensional forms from which most objects can be constructed, the constellation of features that go to form particular objects has to rely on experience and the ultimate process of recognition involves matching to these stored descriptions via their similarity. Likewise, for theories that rely on extraction of two-dimensional features, the central premise again is that these products of information processing are stored and that matching to such stored descriptions underlies recognition. So, in just the same way that we have argued that learning and memory are crucial to olfactory object recognition, not surprisingly a similar role emerges in each of the main theoretical approaches to visual object recognition. Needless to say, this stands in marked contrast to a stimulus-driven theory of olfactory processing, whereby the resultant odor percept is a consequence solely of feature extraction. This ignores both the perceptual ecology of olfaction—to extract meaningful patterns of information from the environment—and the large and growing body of data that implies a central role for experience.

A further point of comparison is conscious access to information at various levels of processing. In the olfactory system a considerable amount of evidence suggests that the underlying features present in an object (equivalent to its constituent chemicals) cannot be accessed directly or that, at best, access is highly constrained. In the visual system there are both similarities to and differences from this situation. The similarity is that visual perception clearly does not include access to some of the processing steps. For example, we cannot readily express the invariances that may govern our perception of objects under different transformations—such information is clearly not accessible. However, a crucial difference is that although practice can only improve the ability to extract feature-based information from a complex olfactory object to a highly limited degree, it clearly can improve the ability to extract information from a complex visual object. One explanation for this difference may be redundancy, in that the sheer size of the processing resources devoted to vision produce a greater amount of information than can be readily used at any one point in time. This would allow for a progressive shift (i.e., perceptual learning) in how attentional resources were allocated to the features of the stimulus. Crudely put, such feature-based information in the visual system may always be present but is typically ignored. In the olfactory system that information may never be available.

A fourth point is that nearly all theorists of visual object recognition acknowledge that object recognition per se is not one thing; rather, it is a set of independent subsystems (Logothetis and Sheinberg 1996). These may include, but are not limited to, a specific face identification module, specific emotion recognition modules, a module for recognizing living and nonliving things, recognition of prototypical members of a specific object category, recognition of subordinate members of a specific object category, and the planning and execution of movements when interacting with commonly encountered objects. Evidence for these various distinctions has arisen in the main from neuropsychological evidence, but experimental evidence from intact individuals also supports at least some of these specific subsystems. In the olfactory system, the object recognition system appears to divide between stimulus-specific entities (notably certain pheromones) and the recognition of complex chemical entities (most olfactory stimuli). There does not appear to be a sharp divide between these different approaches to recognition, as many pheromones turn out not to be single chemicals but, in fact, carefully formulated blends, which may differ only slightly from other pheromones emitted by similar and different species. Thus where the organism is prepared to identify a particular odor object of this sort, this may reflect an innate object recognition system, specific to a particular blend of chemicals or a pattern acquired during some critical period during development. More importantly, the human olfactory neuropsychological evidence only supports a distinction between intensive and qualitative processing, and even this distinction rests on post hoc interpretation of data that was never directly aimed at testing such a distinction. It may turn out that nonhuman species, especially those that devote greater cortical resources to olfaction, do in fact have separate recognition systems and if this is so, perhaps one candidate would be for kin recognition because of its adaptive value in both mate selection and kin identification—the olfactory equivalent perhaps of facial identification. In humans, as we discuss in the penultimate part of this chapter, olfactory neuropsychological research is in its infancy and more refined experimental work in neuropsychological populations may reveal distinctions that are as yet unknown. Furthermore, slow progress in this respect stems from the fact that many patients with potentially interesting olfactory abnormalities are either not routinely tested for olfactory function, fail to report more subtle deficits, or their olfactory problems are simply eclipsed by more severe deficits in other areas. Vision has a clear advantage in that deficits will typically be very apparent and will show up in routine (i.e., visually based) testing.

A further and perhaps the most crucial distinction between visual and olfactory object recognition is the role of language. The relationship between language and olfaction is clearly different from that between vision and language. This linguistic poverty is most readily apparent in the difficulty that participants have in naming even common odors, which is in marked contrast to naming objects encountered in the visual domain. In fact, in visual cognition, a major concern over the years has been the degree to which object recognition is primarily defined by semantic knowledge—language—rather than by the perceptual characteristics of objects. Three lines of research suggest that it is perception that shapes language, rather than the other way round. First, preverbal infants form object categorization schemes that maximize perceptual differences between stimuli, closely paralleling those of verbal children and adults (Eimas and Quinn 1994). Second, categorization of novel objects by adults can occur in the absence of appropriate verbal descriptors and again results in behavior that acts to maximize perceptual differences between items (Gauthier 1995). Third, many animal species also show similar object categorization schemes, in which the most efficient level of discrimination characterizes an object (e.g., Bowman and Sutherland 1970). Thus the human use of language in the description of visual objects reflects the most effective level of discrimination or as it has been termed, the entry-level description (Rosch, Mervis, Gray, Johnson, and Boyes-Braem 1976). So when we see a dog, we typically say "dog," not "mammal" or "beagle." Moreover, the entry-level verbal description tends to follow changes in the perceptual categorization of objects, so that with experts (e.g., plane spotters, dog lovers, or bird aficionados) who demonstrate enhanced perceptual discrimination within their domain of expertise, a change is also observed in their entry-level description of objects within that domain (e.g., a reed warbler rather than a bird).

In considering olfaction, it is interesting to reflect on its naming hierarchy (i.e., entry level, subordinate, and superordinate categories) and the relationship of this to the parallel problem of perceptual expertise. As we have already noted olfactory experts butt up against the same perceptual limit in terms of identifying the components of a complex mixture as do nonexperts. In just the same way, linguistic analysis of olfactory descriptors is very flat, with mainly entry-level terms, but few subordinate (e.g., beagle) or superordinate (e.g., mammal) categories. The fact that language reflects perceptual experience in the visual system might reasonably suggest a similar parallel in the olfactory system. In just the same way that we have argued that olfaction is a synthetic sense, the use of language here appears to reflect this in just the same way that language reflects a more analytic approach within the visual system.

In sum, the key parallel between object recognition systems in the visual and olfactory system is their clear reliance on learning and memory. Undoubtedly, the visual system offers a much more complex picture, in part, because it has had a greater range of demands placed on it, which have in turn favored the evolution of specialist recognition systems. The visual system also enables object recognition to occur at various levels of detail, dependent on perceptual learning and semantic processing. The olfactory system is more constrained in terms of enhanced discriminability of features, but it is just as efficient in discriminating a large range of objects as the visual system, though with same primary reliance on learning and memory.

Do the Olfactory and Visual Systems Employ Similar Basic Neural Circuitries?

Important and profound differences exist between olfaction and other mammalian sensory systems, both in terms of their information processing and the underlying neural circuitry (Koster 2002). However, further examination reveals potentially analogous circuit components in olfaction and vision to deal with perhaps analogous information-processing tasks. As noted above, vision is chosen as the comparator system here because of the wealth of data regarding its structure and function and because of what we interpret as similarities in experience-based object perception in the two systems.

Both vision and olfaction appear to have two distinct modes of sensory discrimination (see chapter 5). One is a largely innate, highly adapted process for detection and discrimination of invariant biologically significant stimuli. Thus, dragonflies, which prey on small flying insects like mosquitoes, have hard-wired circuits that produce neurons with receptive fields consistent with small flying spots. Similarly, male *Manduca* moth olfactory systems have specialized glomerular circuits dedicated to processing female pheromones. Similar feature- or object-detecting circuits have been described in many animals including mammals, although in mammalian vision they may be preferentially expressed early in development (Sewards and Sewards 2002).

The second mode of sensory discrimination in both vision and olfaction, and the one emphasized in this volume, is experience-dependent object synthesis. This mode requires a very different architecture than the hard-wired system, including extensive recurrent and association connections and experience-based plasticity. Both vision and olfaction have basic structural similarities such as (1) small clusters of feature processors organized such that neurons encoding similar features tend to be spatially near each other, (2) feedforward projection pathways that allow convergence of co-occurring features onto individual target neurons, (3) broadly dispersed associative and recurrent pathways expressing synaptic plasticity which enhance feature convergence, and allow for experience-dependent memorization of familiar patterns and formation of perceptual objects, (4) encoding of those perceptual objects by ensembles of neurons in circuits that allow completion of degraded input patterns, and (5) feedback pathways that may allow experience, attentional or expectancy-based modulation of more peripheral stages of information processing.

In the primary visual neocortex, neurons with similar tuning properties, both in terms of spatial location within the visual field and in terms of stimulus orientation preferences, are located near each other in clusters called columnar columns. Columns of neurons expressing slightly different tuning properties are located nearer to each other than columns of neurons with distinctly different tuning properties. In the olfactory bulb, neurons with similar tuning properties are clustered in glomerular columns, with glomeruli tuned to similar molecular features located near to each other (Inaki et al. 2002; Johnson et al. 2002). The metric for similarity/dissimilarity is not as simple (or at least as obvious) with olfactory stimuli as the spatial location or orientation of visual stimuli. Nonetheless, there does appear to be some organization to glomerular patterns, with, for example, glomeruli responsive to aldehydes clustered near each other and glomeruli responsive to alcohols clustered elsewhere.

Furthermore, in the visual cortex, columns with similar tuning properties share reciprocal associative connections that may help synchronize multicolumnar activity (Gilbert and Wiesel 1989). Similarly, in the olfactory bulb, a subclass of tufted cells form intrabulbar associative connections between glomerular columns that receive similar olfactory receptor input (Lodovichi, Belluscio, and Katz 2003). The role of the dual (medial and lateral) representation of the olfactory receptor input to the bulb in odor perception is not known, but these associative connections could facilitate coincident activity in glomerular columns with similar odorant receptive fields, similar to the intercolumnar connections in the primary visual cortex.

In the visual system, features extracted by initial stages of the sensory pathway are actively merged into perceptual objects in the inferotemporal cortex. Faces and complex visual objects are encoded by disparate ensembles of inferotemporal cortical neurons through both anatomical convergence of feature-signaling afferents and experience-dependent plasticity of intracortical association fibers. Although neurons within an inferotemporal cortical column express similar object-receptive fields, the patterning of object representations across the cortex is much less regular than, say, the representation of visual angle in the primary visual cortex. Thus, as in the olfactory bulb, the metric for how information is spatially organized within the inferotemporal cortex is not immediately obvious. This may in part reflect the role of experience in shaping these complex receptive fields in inferotemporal cortex.

In the olfactory system, feature information appears to be merged into perceptual objects within the olfactory cortex, specifically the piriform cortex. As in the inferotemporal visual cortex, object synthesis may occur through both anatomical convergence of cortical afferents from glomerular columns in the olfactory bulb and through experience-dependent plasticity of intracortical association fibers. Given the wide patches of mitral cell termination within the anterior piriform cortex (Zou et al. 2001) and the wider spread of association fiber terminations (Johnson et al. 2000), it might be expected that representation of a given perceptual odor object would be mediated by dispersed ensembles of cortical neurons and that there would be minimal spatial organization of odor object representations. This seems to be supported by recent data (Illig and Haberly 2003).

As noted by Haberly (2001), not only do parallels exist in the sensory physiology of the inferotemporal cortex and the piriform cortex, there are also strong parallels between their extrinsic connections. For example, both the inferotemporal cortex and the piriform cortex have strong reciprocal connections to the amygdala, prefrontal entorhinal, and perirhinal cortices. Thus, in both cases, learned perceptual objects could include complex multimodal, emotional components or at least be capable of evoking multimodal components. The shorter, more direct link between the olfactory system and these structures so heavily involved in memory and emotion, however, may make distinctions between odor percepts and their experiential and emotional associations more difficult to consciously dissect. That is, olfactory representation may have contextual or multimodal features as integral, indivisible components. The more extensive circuitry underlying visual object synthesis, and the inclusion of a spatial dimension to visual objects, may facilitate visual object perceptual dissection compared with odor objects.

While the primary olfactory pathway traditionally extends from the olfactory receptor sheet to the piriform cortex, the piriform cortex projects in turn both directly and indirectly (via the dorsomedial nucleus of the thalamus) to the orbitofrontal cortex. The olfactory system thus has several levels of higher-order cortex, similar to the visual system, presumably each with different functions. One apparent function of the orbitofrontal cortex is convergence with gustatory and perhaps other sensory inputs. Very little sensory physiology has been performed on either the dorsomedial thalamus or the orbitofrontal cortex, although single units in the orbitofrontal cortex do not seem to be drastically different in their responses to odors and context compared to piriform cortex units (Schoenbaum and Eichenbaum 1995). Both the olfactory thalamus and orbitofrontal cortex deserve further attention in future research.

Finally, a critical component of the visual system is feedback. Feedback exists from higher-order cortex to lower order and from cortex to thalamus. Feedback allows past experience, expectancy, and active search processes to highlight some aspects of the visual scene and ignore other aspects as it enters the cortical flow of information. It serves as a form of online, dynamic filtering.

The effect of cortical feedback on olfactory bulb sensory physiology has been seriously understudied, though substantial technical problems may contribute to this weakness. It is clear, however, that mitral cell (Pager 1974; Kay and Laurent 1999) and local field potential (Freeman 2000) activity within the olfactory bulb not only reflect odor stimulation, but also past experience, context, behavioral state, and expectancy. Similar contextual and experience-dependent effects are evident in the activity of the piriform and orbitofrontal cortices (Schoenbaum and Eichenbaum, 1995; Chabaud et al. 2000).

An examination of olfactory system anatomy combined with a comparison with similar foundations in vision thus suggests that olfactory system sensory physiology is not a passive reflection of what enters the nose. Spatiotemporal maps of odorant features expressed in the olfactory bulb are actively read in the light of context and experience. In some situations features A and B are part of a larger synthetic perceptual odor object, whereas in other situations one of the features may be filtered as part of background and the other contributes to an attended to odor object. Experience-dependent cortical synthesis and contextual feedback thus create a flexible system capable of dealing with complex, dynamic stimuli against complex backgrounds — as occurs in both vision and olfaction. A complete understanding of olfactory perception requires far more than simply knowing what odorant molecules and molecular features reach the olfactory epithelium, just as knowing what wavelengths and luminance patterns strike the retina is insufficient to predict what will be visually perceived.

WHAT RESEARCH PROGRAM DOES AN OBJECT RECOGNITION APPROACH SUGGEST?

Experimental Psychology

The study of acquired odor characteristics is in its earliest stages. First, relatively few studies have examined the effect of exposure on discriminability. This is a key point for a mnemonic-based theory of object recognition because the theory suggests that exposure, and hence the encoding of the odor's pattern, should result in enhanced discriminability. Two issues arise here. First, there has recently been concern in the literature over the reliance on nonparametric measures of discriminability such as A' (Pastore et al. 2003). Unfortunately, the most convincing human study of olfactory exposure effects used this statistical technique. Second, some of the published studies appear to manifest improvements in discriminability primarily through a reduction in false-alarm rates, rather than as we might expect an increase in hit rate. Consequently, more detailed studies of exposure effects are needed to ascertain whether they do support the conclusion that we have drawn from them here, namely, that exposure can act to enhance discriminability.

A related issue concerns the parameters of odor-odor learning, that is, the acquisition of odor qualities following exposure to a binary mixture and the subsequent effects on perception of its parts. Although this effect has now been replicated many times, this has only been in one laboratory and, of more empirical concern, is the finding that the effect does not appear to occur for all odor mixtures. As we speculated earlier on, this may result from the relative familiarity of the parts used to make the mixture and/or the role that identifying the mixture elements plays in this process. Counterintuitively, ability to identify the component elements appears to correlate with larger effects, however this may simply be a procedural artifact, in that it may reflect a commonality of terminology between the characteristics identified by the experimenter and the participant. Nonetheless, systematic investigation of the parameters that govern odor-odor learning would clearly be important for determining how the olfactory system binds disparate olfactory experience into an odor object.

The rapidity of globalization, especially of dietary and household goods, should be of major concern to olfactory scientists, because it will, as time advances, make it harder to explore the effects of cultural differences on odor perception. An ideal study might compare the discriminability of odors common and familiar in one culture with a culture where such odors are rarely encountered and vice versa. Yet this type of study will increasingly prove hard to conduct as more of the world's population is exposed to the same type of smells. Several alternative approaches are, however, available and are yet to be explored. First, certain inherited genetic conditions such as phenylketonuria impose a restricted diet on its sufferers from very early on that excludes protein-rich foods. Such individuals might then be poorer in their ability to discriminate the odors of such foods, relative to normal controls. Second, few studies have explored whether discriminative abilities within an olfactory expert's area of expertise (e.g., beer, cheese, tea, etc.) really do exceed those of normal controls, when attention is paid to passive exposure as it recently has been in wine tasting. Third, individuals who raise their children as vegetarians or vegans might also be producing offspring who are selectively poorer at discriminating meat-based smells, relative to those not raised on such a diet. These types of natural experiments could provide interesting insights into the processes underlying object recognition.

The overarching similarity of all the above is the role that passive exposure to odors has on the ability to discriminate between them. One might raise two general comments about this. The first is perhaps the need to develop new approaches to testing olfactory discrimination that enhance both the sensitivity and speed with which such information can be experimentally obtained. Olfactory discrimination using triangle tests or other techniques (duo-trio, 2-alternative forced choice, etc.) are very time consuming and yield relatively little data per participant. Techniques such as those pioneered by Rabin and Cain, which involve the identification of a target odor in a mixture are the sort of thing we have in mind. Second, we are *presuming* that these exposure processes do involve learning, yet in humans the physiology of this process is unknown and whether such learning is prevented by cholinergic antagonists like scopolamine has not been investigated. Because the characteristics of exposure effects, such as their rapidity and resistance to interference, do not resemble standard learning models very well, developing more detailed theoretical accounts that can encompass some of the apparently contradictory findings is needed too.

As we suggested earlier, there are good reasons to suppose that hedonic learning and perceptual learning may have different characteristics. The foundation for this assumption is still rather limited. For example, nobody has as yet directly compared whether the perceptual changes noted in odor-taste learning are as sensitive to counter-conditioning as its hedonic consequences. Similarly, we do not know whether the effect of verbal labels on hedonic judgments of odors reflect the same degree of plasticity, as the effect of verbal labels on an odor's perceptual qualities. In both these cases we might expect greater plasticity for hedonic responding.

The application of cognitive neuroscience approaches, neuroimaging and neuropsychology, to the study of olfaction has a long history, but unfortunately for us its focus has been largely empirically driven, rather than theory based. Neuroimaging offers an exciting avenue to study whether learning processes for hedonic, odor-odor, odor-taste, and mere exposure techniques involve similar or divergent learning systems. Likewise, neuropsychology allows us to test whether the deficits observed in HM, of absent odor quality perception, is something particular to him, or whether this occurs more generally in patients with lesions to the piriform cortex. The discovery of more patients who are unable to differentiate smell on the basis of quality, but who can on the basis of intensity, and who show intact sensitivity on similar tests in other sensory modalities, would be a major boost for an object recognition account of olfactory perception. Moreover the close link observed between qualitative deficits in olfactory perception and dense amnesia also demands exploration to identify whether a common neural substrate may underlie both conditions.

A further question that has seen relatively little exploration in the neuropsychological literature is whether appropriate hedonic judgments for odors can occur in the absence of object recognition. That is, for HM for example, although not able to smell the qualities petrol and coffee, would coffee still smell more pleasant than petrol? We have presumed here that object recognition takes place before the assignment of hedonic tone, yet there is very little evidence to support this assertion. To find, as two papers suggest, that appropriate hedonic responses *may* occur in the absence of object recognition would indeed be a most challenging and fascinating finding, as it would imply two separate olfactory recognition systems in humans.

The constraints that are imposed on a mnemonic system have been quite well explored in the human literature, but not in the animal literature. With two exceptions, we know little about the ability of animals to discriminate component parts of complex mixtures, about the degree to which they can redintegrate a percept from a fragmented stimulus, their ability to engage in exposure-based learning, both unimodal and cross-modally, the extent to which the latter are similarly resistant to interference and the neural substrates that underpin these effects.

Finally, a set of more general issues have been raised by our theoretical perspective. First, exactly how do top-down and bottom-up processing interrelate? We do not know whether top-down processing actually affects perception, in terms of say perceived odor quality or discriminability, or whether any such effects are a consequence of differing descriptions of the same odor. Second, more generally for olfactory cognition, we are still unclear as to whether there are separate and dissociable short- and long-term stores, whether the difficulty that participants have in evoking odor images and the difficulty they have in naming odors are related, and more fundamentally, why odor naming is so poor. Third, we started this chapter by reiterating the differences between a labeled-line system dependent with fixed responses against a plastic system with flexible responses. Exactly how valid is this distinction in the light of the finding that many pheromones, the archetypal example of fixed responding, are themselves odor mixtures of sometimes high complexity?

Sensory Physiology

The view of olfaction outlined in this volume also leads to a variety of specific, testable predictions about the sensory physiology of the mammalian olfactory system. These predictions specifically derive from the view of olfaction as an experience-dependent, object-synthesizing process, and as such should promote novel lines of investigation currently being overlooked. Four such predictions regarding olfactory sensory physiology are outlined here.

First, it is predicted that odor coding and internal representation of odor quality should vary with past experience. These changes could be expressed at any level of the central olfactory system and should outlast duration of the induction stimulus. That is, even during very short stimulus presentations, odor responses of glomeruli, second- and third-order olfactory neurons change. These changes reflect adaptation and dynamic modulation of local circuit function on the scale of seconds. However, there should also be lasting impressions of past odor experience, distinct from these short-term changes. Thus, familiar odorant features may be differentially encoded (e.g., single cell evoked spike rate, odorant receptive field width, or ensemble spike synchrony) compared with the same odorant features when experienced for the first time. Evidence exists for such changes (Fletcher and Wilson 2003), although additional work is needed. How long this effect of familiarity lasts may depend on the duration of past exposure and/or on the nature of the past exposure. If the odorant was experienced in an aversive conditioning context, for example, the extent or duration of change in feature encoding may be enhanced.

The odorant responses of third-order neurons and/or neurons responsible for odor object perception should also reflect odorant familiarity. Odorant discrimination ability or ensemble spiking activity may be expected to change as odor objects are learned. Evidence for such changes in piriform cortical neuron odorant receptive fields has been reported (Wilson 2003). As noted below, the changes in odorant receptive fields may include enhanced robustness when responding to slightly altered inputs, as for example with changes in odorant intensity or variations in odorant backgrounds.

Given that some biologically significant odorants may be encoded by more hardwired circuits (e.g., odors of pheromones or predators), it is predicted that central encoding of some odorants will not be affected by familiarity or past experience. Thus stimulus selection will be important in testing these hypotheses.

A second prediction stemming from the mnemonic view of olfaction is that olfactory bulb glomerular layer odor maps will be read differently depending on past experience and temporal patterns of odor input. For example, presentation of an odor against a background will produce glomerular activation that includes features of both background and foreground odors. How this mixed map is read by cortical circuits should depend on whether the background or foreground odors are familiar. Features of familiar objects will be preferentially grouped into those objects, despite the presence of discordant features that may belong to background, and greater confusion and mixing should occur if the odors are unfamiliar. Figure-ground separation should also be most effective when the background stimulus begins prior to the onset of the figure stimulus, providing an opportunity for cortical circuits to adapt to the background stimulus. These two predictions suggest that how glomerular activity maps are read will vary and suggest that knowing the map alone will not invariably allow prediction of the odor percept.

A third prediction is that odor object synthesis should enhance perceptual constancy despite stimulus intensity induced changes in glomerular odor maps. This is similar to the second prediction in that functional imaging data suggest that, as stimulus intensity increases, glomerular activity maps change, often expanding to include previously silent glomeruli (Leon and Johnson 2003). This expansion presumably reflects the reduced selectivity of olfactory receptor neurons at high stimulus concentrations (Malnic et al. 1999). With familiar odorants, changes in stimulus concentration and thus odor maps should be compensated for by the robust nature of pattern completion and recognition of familiar object synthesis circuitry in the piriform cortex (Barkai et al. 1994). There should be some limits to perceptual constancy, however, wherein odorant that evoked drastically different glomerular activity patterns at different concentrations may evoke different percepts.

Finally, a fourth prediction is that different olfactory functions may emerge at different stages of olfactory system ontogeny. For example, in rodents mitral cells and their targets in the olfactory cortex are present early in development, though their dendritic projections to glomeruli are exuberant and less well defined during the first postnatal week compared with mature rats. Given that these basic convergent projections to cortex and intracortical association fibers are present early, simple odor discrimination (behaviorally and at the single-unit level) might be expected to be relatively normal (Fletcher et al. 2005). However, the primary inhibitory interneuron in the olfactory bulb, granule cells, emerge very late in development. If these cells subserve lateral inhibition, then receptive fields and behavior discrimination of very similar odorants might be expected to be impaired early in development. On the other hand, if the primary function of granule cells instead is to serve as a target of cortical feedback so as to allow expectation and context to shape odor discrimination and recognition, then simple odor discrimination may be intact in young animals. In this latter view, more complex processes such as figure-ground separation, expectation, and contextual modulation of perception may not be expected to occur until later in development.

Regarding development and the effects of experience, it is also important to recognize that most laboratory animals are raised in severely restricted olfactory environments. Olfactory restriction and enrichment during early development can have profound effects on olfactory system structure and function (Brunjes and Frazier 1986). It will be important to consider to what extent such systems truly reflect normalcy, or whether greater care should be taken to provide macrosmatic animals with richer sensory environments.

CONCLUSION

The transduction of electromagnetic energy by the eye and the processing of this information by the brain allow us to detect, discriminate, and recognize a wide variety of objects. These visual objects can be recognized in a variety of different orientations, under different lighting conditions, and when partially occluded. These ecological aspects of vision have been of primary importance in shaping theories of object recognition, all of which rely on learning and memory. In olfaction, a similar focus on the ecology of perception requires a system capable of recognizing entities composed of tens or hundreds of chemicals, against a shifting and equally complex olfactory background. In just the same way that contemporary theories of visual object recognition have been shaped by the demands that the visual environment places on it, a similar acknowledgment in olfaction demands a system that can correlate and encode features and then later extract this information from a complex olfactory scene — object recognition. Similarly, just as object recognition in vision relies, as the name itself implies, on a major role for learning and memory, so too we have argued here that the same applies with equal force in olfaction. If this is not enough, a growing body of experimental evidence, much of which has been reviewed in this book, neurophysiological, neuropsychological, and psychological, also converges on this conclusion. In sum, learning and memory are vital components in a comprehensive understanding of olfactory perception as they underpin the most crucial process in determining what we perceive when smelling an odor, its recognition as a discrete and highly correlated combination of chemicals — an odor object.

Bibliography

Abdi H (2002) What can cognitive psychology and sensory evaluation learn from each other? Food Qual Preference 13:445–451.

Abraham A, Mathai KV (1983) The effect of right temporal lobe lesions on matching of smells. Neuropsychologia 21:277–281.

Abramson CI, Aquino IS, Silva MC, Price JM (1997) Learning in the Africanized honey bee: Apis mellifera L. Physiol Behav 62:657–674.

Adler J (1969) Chemoreceptors in bacteria. Science 166:1588–1597.

Adrian ED (1942) Olfactory reactions in the brain of the hedgehog. J Physiol 100:459–473.

———. (1950) The electrical activity of the mammalian olfactory bulb. Electroencephalogr Clin Neurophysiol 2:377–388.

Aggleton JP, Waskett L (1999) The ability of odours to serve as state-dependent cues for real-world memories: Can Viking smells aid in the recall of Viking experiences? Br J Psychol 90:1–7.

Agster KL, Fortin NJ, Eichenbaum H (2002) The hippocampus and disambiguation of overlapping sequences. J Neurosci 22:5760–5768.

Algom D, Cain WS. (1991) Remembered odors and mental mixtures: Tapping reservoirs of olfactory knowledge. J Exp Psychol Hum Percept Perform 17:1104–1119.

Algom D, Marks LE, Cain WS (1993) Memory psychophysics for chemosensation: Perceptual and mental mixtures of odor and taste. Chem Senses 18:151–160.

Amoore JE (1952) The stereochemical specificities of human olfactory receptors. Perfumery Essential Oil Record 43:321–331.

———. (1970) Molecular basis of odor. Springfield, Ill.: Charles C Thomas.

———. (1975) Four primary odor qualities in man: Experimental evidence and possible significance. In: Olfaction and taste V, Proceedings of the Fifth International Symposium Held at the Howard Florey Institute of Experimental Physiology and Medicine, University of Melbourne, Australia, October 1974, ed. Denton DA, Coghlan JP, 283–289. New York: Academic Press.

——— . (1982) Odor theory and odor classification. In: Fragrance chemistry: The science of sense of smell, ed. Theimer ET, 27–76. New York: Academic Press.

Amoore JE, Forrester LJ (1976) Specific anosmia to trimethylamine: The fishy primary odor. J Chem Ecol 2:49–56.

Anderson AK, Christoff K, Stappen I, Panitz D, Ghahremani DG, Glover G, Gabrieli JD, Sobel N (2003) Dissociated neural representations of intensity and valence in human olfaction. Nat Neurosci 6:196–202.

Annett JM, Leslie JC (1996) Effects of visual and verbal interference on olfactory memory: The role of task complexity. Br J Psychol 87:447–457.

Annett JM, Lorimer AW (1995) Primacy and recency in recognition of odours and recall of odour names. Percept Mot Skills 81:787–794.

Araneda RC, Kini AD, Firestein S (2000) The molecular receptive range of an odorant receptor. Nat Neurosci 3:1248–1255.

Arctander S (1969) Perfume and flavor chemicals. Montclair, NJ: Arctander.

Ashton R, White KD (1980) Sex differences in imagery vividness. An artifact of the test. Br J Psychol 71:35–38.

Atkinson RC, Shiffrin RM (1968) Human memory: A proposed system and its control processes. In: The psychology of learning and motivation, ed. Spence KW, Spence JT. London: Academic Press.

Ayabe-Kanamura S, Kikuchi T, Saito S (1997) Effect of verbal cues on recognition memory and pleasantness evaluation of unfamiliar odors. Percept Mot Skills 85:275–285.

Ayabe-Kanamura S, Schicker I, Laska M, Hudson R, Distel H, Kobayakawa T, Saito S (1998) Differences in perception of everyday odors: A Japanese-German cross cultural study. Chem Senses 23:31–38.

Baddeley AD, Andrade J (2000) Working memory and the vividness of imagery. J Exp Psychol Gen 129:126–145.

Baddeley AD, Hitch GJ (1974) Working memory. In: Recent advances in learning and motivation, ed. Bower GA, 47–89. New York: Academic Press.

Baeyens F, Crombez, G, De Houwer J, Eelen P (1996) No evidence for modulation of evaluative flavor-flavor associations in humans. Learn Motiv 27:200–241.

Baeyens F, Crombez G, Hendrickx H, Eelen P (1995) Parameters of human evaluative flavor-flavor conditioning. Learn Motiv 26:141–160.

Baeyens F, Eelen P, Van den Bergh O, Crombez G (1989) Acquired affective-evaluative value: Conservative but not unchangeable. Behav Res Ther 27:279–287.

Baeyens F, Eelen P, Van den Bergh O, Crombez G (1990) Flavor-flavor and color-flavor conditioning in humans. Learn Motiv 21:434–455.

Baeyens F, Wrzesniewski A, De Houwer J, Eelen P (1996) Toilet rooms, body massages, and smells: Two field studies on human evaluative odor conditioning. Curr Psychol 15:77–96.

Baker CI, Behrmann M, Olson CR (2002) Impact of learning on representation of parts and wholes in monkey inferotemporal cortex. Nat Neurosci 5:1210–1216.

Barkai E, Bergman RE, Horwitz G, Hasselmo ME (1994) Modulation of associative memory function in a biophysical simulation of rat piriform cortex. J Neurophysiol 72:659–677.

Barkai E, Saar D (2001) Cellular correlates of olfactory learning in the rat piriform cortex. Rev Neurosci 12:111–120.

Baron RA, Bronfen MI (1994) A whiff of reality: Empirical evidence concerning the effects of pleasant fragrances on work-related behavior. J Appl Soc Psychol 24:1179–1203.

Bartels A, Zeki S (2000) The neural basis of romantic love. NeuroReport 11:3829–3834.

Barton RA, Purvis A, Harvey PH (1995) Evolutionary radiation of visual and olfactory brain systems in primates, bats and insectivores. Philos Trans R Soc Lond B Biol Sci 348:381–392.

Baydar A, Petrzilka M, Schott MP (1993) Olfactory thresholds for androstenone and Galaxolide: Sensitivity, insensitivity and specific anosmia. Chem Senses 18:661–668.

Beauchamp GK, Wellington JL (1984) Habituation to individual odors occurs following brief, widely-spaced presentations. Physiol Behav 32:511–514.

Beauchamp GK, Yamazaki K, Boyse EA (1988) The chemosensory recognition of genetic individuality. Sci Am 253:86–92.

Beauchamp GK, Yamazaki K, Wysocki CJ, Slotnick BM, Thomas L, Boyse EA (1985) Chemosensory recognition of mouse major histocompatibility types by another species. Proc Natl Acad Sci USA 82:4186–4188.

Beets MGJ (1978) Structure-activity relationships in human chemo-reception. London: Applied Science.

Bensafi M, Porter J, Pouliot S, Mainland J, Johnson B, Zelano C, Young N, Bremner E, Aframian D, Khan R, Sobel N (2003) Olfactomotor activity during imagery mimics that during perception. Nat Neurosci 6:1142–1144.

Berkeley G, Dancy J (1998) A treatise concerning the principles of human knowledge. New York: Oxford University Press.

Bermudez-Rattoni F, Grijalva CV, Kiefer SW, Garcia J (1986) Flavor-illness aversions: the role of the amygdala in the acquisition of taste-potentiated odor aversions. Physiol Behav 38:503–508.

Bernstein IL, Webster MM (1980) Learned taste aversions in humans. Physiol Behav 25:363–366.

Best AR, Thompson JV, Fletcher ML, Wilson DA (2005) Cortical metabotropic glutamate receptors contribute to habituation of a simple odor-evoked behavior. J Neurosci 25:2513–2517.

Best AR, Wilson DA (2004) Coordinate synaptic mechanisms contributing to olfactory cortical adaptation. J Neurosci 24:652–660.

Betts GH (1909) The distribution and functions of mental imagery. New York: Columbia University Teachers College.

Biederman I (1987) Recognition-by-components: A theory of human image understanding. Psychol Rev 94:115–147.

Biederman I, Gerhardstein PC (1993) Recognizing depth-rotated objects: Evidence and conditions for three-dimensional viewpoint invariance. J Exp Psychol Hum Percept Perform 19:1162–1182.

Bilko A, Altbacker V, Hudson R (1994) Transmission of food preference in the rabbit: the means of information transfer. Physiol Behav 56:907–912.

Birch LL, Marlin DW (1982) I don't like it; I never tried it: Effects of exposure on two-year-old children's food preferences. Appetite J Intake Res 3:353–360.

Bodyak N, Slotnick B (1999) Performance of mice in an automated olfactometer: Odor detection, discrimination and odor memory. Chem Senses 24:637–645.

Bolger EM, Titchener EB (1907) Some experiments on the associative power of smells. Am J Psychol 18:326–327.

Booth DA, Mather P, Fuller J (1982) Starch content of ordinary foods associatively conditioning human appetite and satiation, indexed by intake and eating pleasantness of starch-paired flavours. Appetite J Intake Res 3:163–184.

Boring EG (1928) A new system for the classification of odors. Am J Psychol 40:345–349.

Boring EG (1950) A history of experimental psychology. New York: Appleton-Century-Crofts.

Bouret S, Sara SJ (2002) Locus coeruleus activation modulates firing rate and temporal organization of odour-induced single-cell responses in rat piriform cortex. Eur J Neurosci 16:2371–2382.

Bower RL, Doran TP, Edles PA, May K (1994) Paired associate learning with visual and olfactory cues: Effects of temporal order. Psychol Rep 44:501–507.

Bowman RS, Sutherland NS (1970) Shape discrimination by goldfish: coding of irregularities. J Comp Physiol Psychol 72:90–97.

Brand GM (2002) Sex differences in human olfaction: Between evidence and enigma. Q J Exp Psychol B 54:259–270.

Bregman AS (1990) The auditory scene. In: Auditory scene analysis, 1–45. Cambridge, MA: MIT Press.

Bremmer EA, Mainland JD, Khan RM, Sobel N (2003) The prevalence of androstenone anosmia. Chem Senses 28:423–432.

Brennan PA, Keverne EB (1997) Neural mechanisms of mammalian olfactory learning. Prog Neurobiol 51:457–481.

Brud WS (1986) Words versus odours: How perfumers communicate. Perfurmer Flavorist 11:27–44.

Brunjes PC, Frazier LL (1986) Maturation and plasticity in the olfactory system of vertebrates. Brain Res 396:1–45.

Buck L, Axel R (1991) A novel multigene family may encode odorant receptors: A molecular basis for odor recognition. Cell 65:175–187.

Buck LB (1996) Information coding in the vertebrate olfactory system. Annu Rev Neurosci 19:517–544.

Bunsey M, Eichenbaum H (1993) Critical role of the parahippocampal region for paired-associate learning in rats. Behav Neurosci 107:740–747.

Burman OH, Mendl M (2002) Recognition of conspecific odors by laboratory rats (Rattus norvegicus) does not show context specificity. J Comp Psychol 116:247–252.

Burstein A (1987) Olfactory hallucinations. Hosp Community Psychiatry 38:80.

Cabanac M (1971) Physiological role of pleasure. A stimulus can feel pleasant or unpleasant depending upon its usefulness as determined by internal signals. Science 173:1103–1107.

Cain WS (1979) To know with the nose: Keys to odor and identification. Science 203:468–470.

———. (1982) Odor identification by males and females: Predictions vs performance. Chem Senses 7:129–142.

Cain WS, Algom D (1997) Perceptual and mental mixtures in odor and in taste: Are there similarities and differences between experiments or between modalities? Reply to Schifferstein (1997) J Exp Psychol Hum Percept Perform 23:1588–1593.

Cain WS, Gent JF (1991) Olfactory sensitivity: Reliability, generality, and association with aging. J Exp Psychol Hum Percept Perform 17:382–391.

Cain WS, Johnson F (1978) Lability of odor pleasantness: Influence of mere exposure. Perception 7:459–465.

Cain WS, Polak EH (1992) Olfactory adaptation as an aspect of odor similarity. Chem Senses 17:481–491.

Cain WS, Potts BC (1996) Switch and bait: Probing the discriminative basis of odor identification via recognition memory. Chem Senses 21:35–44.

Cain WS, Stevens JC, Nickou CM, Giles A, Johnston I, Garcia-Medina MR (1995) Life-span development of odor identification, learning, and olfactory sensitivity. Perception 24:1457–1472.

Callahan CD, Hinkebein J (1999) Neuropsychological significance of anosmia following traumatic brain injury. J Head Trauma Rehabil 14:581–587.

Cann A, Ross DA (1989) Olfactory stimuli as contextual cues in human memory. Am J Psychol 102:91–102.

Cannon DS, Best MR, Batson JD, Feldman M (1983) Taste familiarity and apomorphine-induced taste aversions in humans. Behav Res Ther 21:669–673.

Capaldi ED (1996) Conditioned food preferences. In: Why we eat what we eat: The psychology of eating, ed. Capaldi ED, 53–80. Washington, DC: American Psychological Association.

Cardello AV, Maller O, Masor HB, Dubose C, Edelman B (1985) Role of consumer expectancies in the acceptance of novel foods. J Food Sci 50:1707–1714.

Carrasco M, Ridout JB (1993) Olfactory perception and olfactory imagery: A multidimensional analysis. J Exp Psychol Hum Percept Perform 19:287–301.

Carroll B, Richardson JT, Thompson P (1993) Olfactory information processing and temporal lobe epilepsy. Brain Cogn 22:230–243.

Case TI, Stevenson RJ, Dempsey RA (2004) Reduced discriminability following perceptual learning with odours. Perception 33:113–119.

Chabaud P, Ravel N, Wilson DA, Mouly AM, Vigouroux M, Farget V, Gervais R (2000) Exposure to behaviourally relevant odour reveals differential characteristics in rat central olfactory pathways as studied through oscillatory activities. Chem Senses 25:561–573.

Chastrette M (1997) Trends in structure-odor relationships. SAR QSAR Environ Res 6:215–254.

———. (2002) Classification of odors and structure-odor relationships. In: Olfaction, taste, and cognition, ed. Rouby C, Schaal B, Dubois D, Gervais R, Holley A, 100–116. Cambridge, U.K.: Cambridge University Press.

Chastrette M, Elmouaffek A, Sauvegrain P (1988) A multidimensional statistical study of 74 notes used in perfumery. Chem Senses 13:295–305.

Chen Y, Getchell TV, Sparks DL, Getchell ML (1993) Patterns of adrenergic and peptidergic innervation in human olfactory mucosa: age-related trends. J Comp Neurol 334:104–116.

Christensen TA, Lei H, Hildebrand JG (2003) Coordination of central odor representations through transient, non-oscillatory synchronization of glomerular output neurons. Proc Natl Acad Sci USA 100:11076–11081.

Christensen TA, Pawlowski VM, Lei H, Hildebrand JG (2000) Multi-unit recordings reveal context-dependent modulation of synchrony in odor-specific neural ensembles. Nat Neurosci 3:927–931.

Chu S, Downes JJ (2000) Long live Proust: The odour-cued autobiographical memory bump. Cognition 75:B41–B50.

Clark CC, Lawless HT (1994) Limiting response alternatives in time-intensity scaling: An examination of the halo dumping effect. Chem Senses 19:583–594.

Clark RE, Manns JR, Squire LR (2002) Classical conditioning, awareness, and brain systems. Trends Cogn Sci 6:524–531.

Classen C (1992) The odor of the other: Olfactory symbolism and cultural categories. Ethos 20:133–166.

Cleland TA, Morse A, Yue EL, Linster C (2002) Behavioral models of odor similarity. Behav Neurosci 116:222–231.

Cleland TA, Narla VA (2003) Intensity modulation of olfactory acuity. Behav Neurosci 117:1434–1440.

Cliff M, Noble AC (1990) Time-intensity evaluation of sweetness and fruitiness and their interaction in a model solution. J Food Sci 55:450–454.

Cohen AB, Johnston RE, Kwon A (2001) How golden hamsters (Mesocricetus auratus) discriminate top from bottom flank scents in over-marks. J Comp Psychol 115:241–247.

Conway MA, Pleydell-Pearce CW (2000) The construction of autobiographical memories in the self-memory system. Psychol Rev 107:261–288.

Costanzo RM, Zasler ND (1991) Head trauma. In: Smell and taste in health and disease, ed. Getchell TV, Doty RL, Bartoshuk LM, Snow JB, 711–730. New York: Raven Press.

Coureaud G, Langlois D, Sicard G, Schaal B (2004) Newborn rabbit responsiveness to the mammary pheromone is concentration-dependent. Chem Senses 29:341–350.

Coureaud G, Schaal B, Langlois D, Perrier G (2001) Orientation response of newborn rabbits to odours of lactating females: relative effectiveness of surface and milk cues. Anim Behav 61:153–162.

Cousens S, Kanki B, Toure S, Diallo I, Curtis V (1996) Reactivity and repeatability of hygiene behaviour: structured observations from Burkina Faso. Soc Sci Med 43:1299–1308.

Cowan N (1988) Evolving conceptions of memory storage, selective attention, and their mutual constraints within the human information-processing system. Psychol Bull 104:163–191.

Cowley JJ, Brooksbank BWL (1991) Human exposure to putative pheromones and changes in aspects of social behaviour. J Steroid Biochem Mol Biol 39:647–659.

Crandall CS (1984) The liking of foods as a result of exposure: Eating doughnuts in Alaska. J Soc Psychol 125:187–194.

Crick F (1984) Function of the thalamic reticular complex: the searchlight hypothesis. Proc Natl Acad Sci USA 81:4586–4590.

Crist RE, Li W, Gilbert CD (2001) Learning to see: experience and attention in primary visual cortex. Nat Neurosci 4:519–525.

Critchley HD, Rolls ET (1996) Olfactory neuronal responses in the primate orbito-frontal cortex: Analysis in an olfactory discrimination task. J Neurophysiol 75:1659–1672.

Crocker EC (1947) Odor in flavor. Am Perfumer Essential Oil Rev 12:164–165.

Crocker EC, Henderson LF (1927) Analysis and classification of odors. Am Perfumer Essential Oil Rev 22:325–327.

Crosley CJ, Dhamoon S (1983) Migrainous olfactory aura in a family. Arch Neurol 40:459.

Crowder RG, Schab FR (1995) Imagery for odors. In: Memory for odors, ed. Schab FR, Crowder RG, 93–107. Mahwah, NJ: Lawrence Erlbaum.

Curtis V, Biran A (2001) Dirt, disgust and disease: Is hygiene in our genes? Perspect Biol Med 44:17–31.

Cutler WB, Friedmann E, McCoy NL (1998) Pheromonal influences on the sociosexual behavior of men. Arch Sex Behav 27:1–13.

Dade LA, Zatorre RJ, Jones-Gotman M (2002) Olfactory learning: Convergent findings from lesion and brain imaging studies in humans. Brain 125:86–101.

Dalton P (2000) Psychophysical and behavioral characteristics of olfactory adaptation. Chem Senses 25:487–492.

Dalton P, Wysocki CJ (1996) The nature and duration of adaptation following long-term odor exposure. Percept Psychophys 58:781–792.

Daly CD, White RS (1930) Psychic reactions to olfactory stimuli. Br J Psychol 10:70–87.

Daly D (1958) Uncinate fits. Neurology 8:250–260.

Datiche F, Roullet F, Cattarelli M (2001) Expression of Fos in the piriform cortex after acquisition of olfactory learning: an immunohistochemical study in the rat. Brain Res Bull 55:95–99.

Davidson RJ, Irwin W (1999) The functional neuroanatomy of emotion and affective style. Trends Cogn Sci 3:11–21.

Davis RG (1975) Acquisition of verbal associations to olfactory stimuli of varying familiarity and to abstract visual stimuli. J Exp Psychol Hum Learn Mem 1:134–142.

———. (1977) Acquisition and retention of verbal associations to olfactory and abstract visual stimuli of varying similarity. J Exp Psychol Hum Learn Mem 3:37–51.

Davis RG, Pangborn RM (1985) Odor pleasantness judgments compared among samples from 20 nations using microfragrances. Chem Senses 30:413.

de Araujo IE, Rolls ET, Kringelbach ML, McGlone F, Phillips N (2003) Taste-olfactory convergence, and the representation of the pleasantness of flavour, in the human brain. Eur J Neurosci 18:2059–2068.

Deems DA, Doty RL, Settle G, Moore-gillon V, Shaman P, Mester AF, Kimmelman CP, Brightman VJ, Snow JB (1991) Smell and taste disorders, a study of 750 patients from the University of Pennsylvania smell and taste center. Arch Otolaryngol Head Neck Surg 117:519–528.

De Houwer J, Thomas S, Baeyens F (2001) Associative learning of likes and dislikes: A review of 25 years of research on human evaluative conditioning. Psychol Bull 127:853–869.

Dempsey RA, Stevenson RJ (2002) Gender differences in the retention of Swahili names for unfamiliar odors. Chem Senses 27:681–689.

Derby CD, Hutson M, Livermore BA, Lynn WH (1996) Generalization among related complex odorant mixtures and their components: analysis of olfactory perception in the spiny lobster. Physiol Behav 60:87–95.

De Rosa E, Hasselmo ME (2000) Muscarinic cholinergic neuromodulation reduces proactive interference between stored odor memories during associative learning in rats. Behav Neurosci 114:32–41.

Desor JA, Beauchamp GK (1974) The human capacity to transmit olfactory information. Percept Psychophys 16:551–556.

De Wijk RA, Cain WS (1994) Odor identification by name and by edibility: Life-span development and safety. Hum Factors 36:182–187.

Dielenberg RA, McGregor IS (2001) Defensive behavior in rats towards predatory odors: A review. Neurosci Biobehav Rev 25:597–609.

Distel H, Ayabe-Kanamura S, Schicker I, Martinez-Gomez M, Kobayakawa T, Saito S, Hudson R (1999) Perception of everyday odors — Correlation between intensity, familiarity and strength of hedonic judgment. Chem Senses 24:191–199.

Distel H, Hudson R (2001) Judgment of odor intensity is influenced by subjects knowledge of the odor source. Chem Senses 26:247–251.

Djordjevic J, Zatorre R, Jones-Gotman M (2004a) Effects of perceived and imagined odors on taste detection. Chem Senses 29:199–208.

Djordjevic J, Zatorre R, Petrides M, Jones-Gotman M (2004) The mind's nose: Effects of odor and visual imagery on odor detection. Psychol Sci 15:143–148.

Dorries KM, Schmidt HJ, Beauchamp GK, Wysocki CJ (1989) Changes in sensitivity to the odor of androstenone during adolescence. Dev Psychobiol 22:423–435.

Doty RL (1997) Studies of human olfaction from the University of Pennsylvania smell and taste center. Chem Senses 22:565–586.

Doty RL, Ford M, Preti G, Huggins GR (1975) Changes in the intensity and pleasantness of human vaginal odors during the menstrual cycle. Science 190:1316–1318.

Doty RL, Reyes PF, Gregor T (1985) The primates III: Humans. In: Social odours in mammals, ed. Brown RE, Macdonald DW, 804–832. Oxford: Clarendon Press.

———. (1987) Presence of both odor identification and detection deficits in Alzheimer's disease. Brain Res Bull 18:597–600.

Doty RL, Shaman P, Applebaum SL, Giberson R, Sikorski L, Rosenberg L (1984) Smell identification ability: Changes with age. Science 226:1441–1443.

Dravnieks A, Bock FC, Powers JJ, Tibbetts M, Ford M (1978) Comparisons of odors directly and through profiling. Chem Senses Flavor 3:191–225.

———. (1986) Atlas of odor character profiles. Philadelphia: American Society for Testing and Materials (ASTM).

Dubnau J, Tully T (1998) Gene discovery in Drosophila: New insights for learning and memory. Annu Rev Neurosci 21:407–444.

Dubose CN, Cardello AV, Maller O (1980) Effects of colorants and flavorants on identification, perceived flavor intensity, and hedonic quality of fruit-flavored beverage and cake. J Food Sci 45:1393–1415.

Duclaux R, Feisthauer J, Cabanac M (1973) Effets du repas sur l'agrement d'odeurs alimentaires et nonalimentaires chez l'homme. Physiol Behav 10:1029–1033.

Dudchenko PA, Wood ER, Eichenbaum H (2000) Neurotoxic hippocampal lesions have no effect on odor span and little effect on odor recognition memory but produce significant impairments on spatial span, recognition, and alternation. J Neurosci 20:2964–2977.

Dusenbery DB (1992) Sensory ecology. New York: W.H. Freeman and Company.

Efron R (1956) The effect of olfactory stimuli in arresting uncinate fits. Brain 79:267–281.

Ehman KD, Scott E (2001) Urinary odour preference of MHC congenic female mice, *Mus domesticus*: Implications for kin recognition and detection of parasitized males. Anim Behav 62:781–789.

Eichenbaum H (1998) Using olfaction to study memory. Ann N Y Acad Sci 855:657–669.

———. (2000) A cortical-hippocampal system for declarative memory. Nat Rev Neurosci 1:41–50.

Eichenbaum H, Fagan A, Mathews P, Cohen NJ (1988) Hippocampal system dysfunction and odor discrimination learning in rats: impairment or facilitation depending on representational demands. Behav Neurosci 102:331–339.

Eichenbaum H, Morton TH, Potter H, Corkin S (1983) Selective olfactory deficits in case H.M. Brain 106:459–472.

Eimas PD, Quinn PC (1994) Studies on the formation of perceptually based basic-level categories in young infants. Child Dev 65:903–917.

Eisthen HL, Delay RJ, Wirsig-Wiechmann CR, Dionne VE (2000) Neuromodulatory effects of gonadotropin releasing hormone on olfactory receptor neurons. J Neurosci 20:3947–3955.

Elmes DG (1998) Is there an inner nose? Chem Senses 23:443–445.

Embril JA, Camfield P, Artsob H, Chase DP (1983) Powassan virus encephalitis resembling herpes simplex encephalitis. Arch Intern Med 143:341–343.

Engen T (1972) The effect of expectation on judgments of odor. Acta Psychol 36:450–458.

———. (1974) Method and theory in the study of odor preferences. In: Human responses to environmental odors, ed. Johnston JW, Moulton DG, Turk A, 121–141. New York: Academic Press.

———. (1982) The perception of odors. New York: Academic Press.

———. (1987) Remembering odors and their names. Am Scientist 75:497–503.

———. (1991) Odor memory. In: Odor sensation and memory, 77–87. New York: Praeger.

Engen T, Kuisma JE, Eimas PD (1973) Short-term memory of odors. J Exp Psychol 99:222–225.

Engen T, Pfaffmann C (1960) Absolute judgments of odor quality. J Exp Psychol 59:214–219.

Engen T, Ross BM (1973) Long-term memory of odors with and without verbal descriptions. J Exp Psychol 100:221–227.

Ennis M, Linster C, Aroniadou-Anderjaska V, Ciombor K, Shipley MT (1998) Glutamate and synaptic plasticity at mammalian primary olfactory synapses. Ann N Y Acad Sci 855:457–466.

Eskenazi B, Cain WS, Novelly RA, Friend KB (1983) Olfactory functioning in temporal lobectomy patients. Neuropsychologia 21:365–374.

Faber T, Joerges J, Menzel R (1999) Associative learning modifies neural representations of odors in the insect brain. Nat Neurosci 2:74–78.

Fallon AE, Rozin P (1983) The psychological bases of food rejections by humans. Ecol Food Nutr 13:15–26.

Farah MJ (2000) The cognitive neuroscience of vision. Malden, MA: Blackwell.

Farooqui T, Robinson K, Vaessin H, Smith BH (2003) Modulation of early olfactory processing by an octopaminergic reinforcement pathway in the honeybee. J Neurosci 23:5370–5380.

Feldman DE, Knudsen EI (1997) An anatomical basis for visual calibration of the auditory space map in the barn owl's midbrain. J Neurosci 17:6820–6837.

Findley AE (1924) Further studies of Henning's system of olfactory qualities. Am J Psychol 35:436–445.

Fleming AS, Corter C, Franks P, Surbey M, Schneider B, Steiner M (1993) Postpartum factors related to mothers attraction to newborn infant odors. Dev Psycobiol 26:115–132.

Fletcher ML, Smith AM, Best AR, Wilson DA (2005) High-frequency oscillations are not necessary for simple olfactory discriminations in young rats. J Neurosci 25:792–798.

Fletcher ML, Wilson DA (2002) Experience modifies olfactory acuity: acetylcholine-dependent learning decreases behavioral generalization between similar odorants. J Neurosci 22:RC201.

———, ———. (2003) Olfactory bulb mitral-tufted cell plasticity: odorant-specific tuning reflects previous odorant exposure. J Neurosci 23:6946–6955.

Frank RA, Byram J (1988) Taste-smell interactions are tasteant and odorant dependent. Chem Senses 13:445–455.

Freeman WJ (1978) Spatial properties of an EEG event in the olfactory bulb and cortex. Electroencephalogr Clin Neurophysiol 44:586–605.

———. (2000) Neurodynamics: An exploration in mesoscopic brain dynamics. London: Springer-Verlag.

Freeman WJ, Baird B (1987) Relation of olfactory EEG to behavior: spatial analysis. Behav Neurosci 101:393–408.

Freeman WJ, Schneider W (1982) Changes in spatial patterns of rabbit olfactory EEG with conditioning to odors. Psychophysiology 19:44–56.

Fried HU, Fuss SH, Korsching SI (2002) Selective imaging of presynaptic activity in the mouse olfactory bulb shows concentration and structure dependence of odor responses in identified glomeruli. Proc Natl Acad Sci USA 99:3222–3227.

Friedrich RW, Laurent G (2001) Dynamic optimization of odor representations by slow temporal patterning of mitral cell activity. Science 291:889–894.

Galef BG Jr, Giraldeau LA (2001) Social influences on foraging in vertebrates: Causal mechanisms and adaptive functions. Anim Behav 61:3–15.

Gauthier I (1995) Becoming a 'greeble' expert: exploring the face recognition mechanism. Masters thesis, Yale University.

Gibson EL, Wainwright CJ, Booth DA (1995) Disguised protein in lunch after low-protein breakfast conditions food-flavor preferences dependent on recent lack of protein intake. Physiol Behav 58:363–371.

Gilbert AN, Crouch M, Kemp SE (1998) Olfactory and visual mental imagery. J Ment Imagery 22:137–146.

Gilbert AN, Wysocki CJ (1987) The smell survey. J Natl Geogr Soc 172:514–525.

Gilbert CD, Wiesel TN (1989) Columnar specificity of intrinsic horizontal and cortico-cortical connections in cat visual cortex. J Neurosci 9:2432–2442.

Glinwood RT, Du YJ, Powell W (1999) Responses to aphid sex pheromones by the pea aphid parasitoids Aphidius ervia and Aphidius eadyi. Entomol Exp Appl 92:227–232.

Goldman WP, Seamon JG (1992) Very long-term memory for odors: Retention of odor-name associations. Am J Psychol 105:549–563.

Gottfried JA, Dolan RJ (2003) The nose smells what the eye sees: Crossmodal visual facilitation of human olfactory perception. Neuron 39:375–386.

Gottfried JA, O'Doherty J, Dolan RJ (2002) Appetitive and aversive olfactory learning in humans studied using event-related functional magnetic resonance imaging. J Neurosci 22:10829–10837.

Goyert H, Frank ME, Gent JF, Hettinger TP (2005) Adaptation and identifying components of olfactory mixtures in humans. Association for Chemoreception Sciences Annual Meeting Abstracts, Sarasota, FL.

Grajski KA, Freeman WJ (1989) Spatial EEG correlates of nonassociative and associative olfactory learning in rabbits. Behav Neurosci 103:790–804.

Grasso FW, Basil JA (2002) How lobsters, crayfishes, and crabs locate sources of odor: current perspectives and future directions. Curr Opin Neurobiol 12:721–727.

Gray CM, Freeman WJ, Skinner JE (1986) Chemical dependencies of learning in the rabbit olfactory bulb: acquisition of the transient spatial pattern change depends on norepinephrine. Behav Neurosci 100:585–596.

Greene J (1986) Language understanding: A cognitive approach. Milton Keynes, U.K.: Open University Press.

Gregory R (1966) Eye and brain: The psychology of seeing. New York: McGraw–Hill.

Gross-Isseroff R, Lancet D (1988) Concentration-dependent changes of perceived odor quality. Chem Senses 13:191–204.

Guillery RW, Feig SL, Lozsadi DA (1998) Paying attention to the thalamic reticular nucleus. Trends Neurosci 21:28–32.

Guillot M (1948) Anosmies partielles et odeurs fondamentales. C R Acad Sci 226:1307–1309.

Haberly LB (1985) Neuronal circuitry in olfactory cortex. Chem Senses 10:219–238.

——. (2001) Parallel-distributed processing in olfactory cortex: new insights from morphological and physiological analysis of neuronal circuitry. Chem Senses 26:551–576.

Haller R, Rummel C, Henneberg S, Pollmer U, Koester EP (1999) The influence of early experience with vanillin on food preference later in life. Chem Senses 24:465–467.

Hamrick WD, Wilson DA, Sullivan RM (1993) Neural correlates of memory for odor detection conditioning in adult rats. Neurosci Lett 163:36–40.

Harley HE, Roitblat HL, Nachtigall PE (1996) Object representation in the bottlenose dolphin (Tursiops truncatus): integration of visual and echoic information. J Exp Psychol Anim Behav Process 22:164–174.

Harper R, Bate-Smith EC, Lad DG (1968) Odour description and odour classification. London: Churchill.

Harris JA, Shand FL, Carroll LQ, Westbrook RF (2004) Persistence of preference for a flavor presented in simultaneous compound with sucrose. J Exp Psychol Anim Behav Process 30:177–189.

Hartlieb E, Anderson P, Hansson BS (1999) Appetitive learning of odours with different behavioural meaning in moths. Physiol Behav 67:671–677.

Hasselmo ME, Barkai E (1995) Cholinergic modulation of activity-dependent synaptic plasticity in the piriform cortex and associative memory function in a network biophysical simulation. J Neurosci 15:6592–6604.

Hasselmo ME, Wilson MA, Anderson BP, Bower JM (1990) Associative memory function in piriform (olfactory) cortex: computational modeling and neuropharmacology. Cold Spring Harb Symp Quant Biol 55:599–610.

Hayar A, Karnup S, Ennis M, Shipley MT (2004) External tufted cells: a major excitatory element that coordinates glomerular activity. J Neurosci 24:6676–6685.

Hays WST (2003) Human pheromones: have they been demonstrated? Behav Ecol Sociobiol 54:89–97.

Hazzard FW (1930) A descriptive account of odors. J Exp Psychol 13:297–331.

Hebb DO (1949) The organization of behavior: A neuropsychological theory. New York: Wiley.

Hedrick PW, Black FL (1997) HLA and mate choice: No evidence in South Amerindians. Am J Hum Genet 61:505–511.

Heinbockel T, Christensen TA, Hildebrand JG (1999) Temporal tuning of odor responses in pheromone-responsive projection neurons in the brain of the sphinx moth Manduca sexta. J Comp Neurol 409:1–12.

Held R (1989) Perception and its neuronal mechanisms. Cognition 33:139–154.

Hellstern F, Malaka R, Hammer M (1998) Backward inhibitory learning in honeybees: a behavioral analysis of reinforcement processing. Learn Mem 4:429–444.

Henkin RI, Levy LM (2002) Functional MRI of congenital hyposmia: Brain activation to odors and imagination of odors and tastes. J Comput Assist Tomogr 26:39–61.

Henning H (1916) Der Geruch. Leipzig, Germany: Barth.

Hepper PG, Waldman B (1992) Embryonic olfactory learning in frogs. Q J Exp Psychol B 44:179–197.

Herz RS (1998) Are odors the best cues to memory? A cross-modal comparison of associative memory stimuli. Ann N Y Acad Sci 855:670–674.

———. (2000) Verbal coding in olfactory versus nonolfactory cognition. Mem Cognit 28:957–964.

Herz RS, Cahill ED (1997) Differential use of sensory information in sexual behavior as a function of gender. Hum Nat 8:275–286.

Herz RS, Cupchick GC (1995) The emotional distinctiveness of odor-evoked memories. Chem Senses 20:517–528.

Herz RS, Engen T (1996) Odor memory: Review and analysis. Psychon Bull Rev 3:300–313.

Herz RS, Schooler JW (2002) A naturalistic study of autobiographical memories evoked by olfactory and visual cues: Testing the Proustian hypothesis. Am J Psychol 115:21–32.

Herz RS, von Clef J (2001) The influence of verbal labeling on the perception of odors: Evidence for olfactory illusions? Perception 30:381–391.

Heywood A, Vortried HA, Washburn MF (1905) Minor studies from the psychological laboratory of Vassar College: Some experiments on the association power of smells. Am J Psychol 16:537–541.

Hildebrand JG, Shepherd GM (1997) Mechanisms of olfactory discrimination: converging evidence for common principles across phyla. Annu Rev Neurosci 20:595–631.

Holldobler B (1995) The chemistry of social regulation: Multicomponent signals in ant societies. Proc Natl Acad Sci USA 92:19–22.

Hudson R (1999) From molecule to mind: the role of experience in shaping olfactory function. J Comp Physiol A 185:297–304.

Hudson R, Distel H (2002) The individuality of odor perception. In: Olfaction, taste, and cognition, ed. Rouby C, Schaal B, Dubois D, Gervais R, Holley A, 408–420. Cambridge: Cambridge University Press.

Hvastja L, Zanuttini L (1989) Odour memory and odour hedonics in children. Perception 18:391–396.

Illig KR, Haberly LB (2003) Odor-evoked activity is spatially distributed in piriform cortex. J Comp Neurol 457:361–373.

Inaki K, Takahashi YK, Nagayama S, Mori K (2002) Molecular-feature domains with posterodorsal-anteroventral polarity in the symmetrical sensory maps of the mouse olfactory bulb: mapping of odourant-induced Zif268 expression. Eur J Neurosci 15:1563–1574.

James W (1890) The principles of psychology. New York: Henry Holt and Company.

Jehl C, Murphy C (1999) Developmental effects on odor learning and memory in children. Ann N Y Acad Sci 855:632–634.

Jehl C, Royet JP, Holley A (1994) Very short term memory recognition for odors. Percept Psychophys 56:658–668.

———. (1995) Odor discrimination and recognition memory as a function of familiarization. Percept Psychophys 57:1002–1011.

———. (1997) Role of verbal encoding in short-and long-term odor recognition. Percept Psychophys 59:100–110.

Jellinek JS, Koster EP (1979) Perceived fragrance complexity and its relation to familiarity and pleasantness. J Soc Cosmet Chem 30:253–262.

Jiang M, Griff ER, Ennis M, Zimmer LA, Shipley MT (1996) Activation of locus coeruleus enhances the responses of olfactory bulb mitral cells to weak olfactory nerve input. J Neurosci 16:6319–6329.

Jinks A, Laing DG (1999) Temporal processing reveals a mechanism for limiting the capacity of humans to analyze odor mixtures. Cogn Brain Res 8:311–325.

———. (2002) The analysis of odor mixtures by humans: evidence for a configurational process. Physiol Behav 72:51–63.

Jinks A, Laing DG, Hutchinson I, Oram N (1998) Temporal processing of odor mixtures reveals that identification of components takes precedence over temporal information in olfactory memory. Ann N Y Acad Sci 855:834–836.

Johnson BA, Ho SL, Xu Z, Yihan JS, Yip S, Hingco EE, Leon M (2002) Functional mapping of the rat olfactory bulb using diverse odorants reveals modular responses to functional groups and hydrocarbon structural features. J Comp Neurol 449:180–194.

Johnson BA, Leon M (2000) Modular representations of odorants in the glomerular layer of the rat olfactory bulb and the effects of stimulus concentration. J Comp Neurol 422:496–509.

Johnson BA, Woo CC, Leon M (1998) Spatial coding of odorant features in the glomerular layer of the rat olfactory bulb. J Comp Neurol 393:457–471.

Johnson BN, Mainland JD, Sobel N (2003) Rapid olfactory processing implicates subcortical control of an olfactomotor system. J Neurophysiol 90:1084–1094.

Johnson DM, Illig KR, Behan M, Haberly LB (2000) New features of connectivity in piriform cortex visualized by intracellular injection of pyramidal cells suggest that "primary" olfactory cortex functions like "association" cortex in other sensory systems. J Neurosci 20:6974–6982.

Johnston RE, Bullock TA (2000) Individual recognition by use of odours in golden hamsters: The nature of individual representations. Anim Behav 61:545–557.

Jones FN, Roberts K, Holman EW (1978) Similarity judgments and recognition in memory for some common species. Percept Psychophys 24:2–6.

Jones MO (2000) What's disgusting, why and what does it matter? J Folklore Res 37:53–72.

Jones-Gotman M, Zatorre RJ (1988) Olfactory identification deficits in patients with focal cerebral excision. Neuropsychologia 26:387–400.

———. (1993) Odor recognition memory in humans: Role of right temporal and orbitofrontal regions. Brain Cogn 22:182–198.

Kadohisa M, Wilson DA (forthcoming) A cortical high-pass filter contributes to olfactory figure-ground separation. J Neurophysiol.

Kalmus H (1955) The discrimination by the nose of the dog of individual human odours and in particular of the odours of twins. Anim Behav 3:25–31.

Kalogerakis MG (1963) The role of olfaction in sexual development. Psychosom Med 25:420–432.

Kareken DA, Mosnik DM, Doty RL, Dzemidzic M, Hutchins GD (2003) Functional anatomy of human odor sensation, discrimination, and identification in health and aging. Neuropsychology 17:482–495.

Kastner S, Ungerleider LG (2000) Mechanisms of visual attention in the human cortex. Annu Rev Neurosci 23:315–341.

Kay LM, Lancaster LR, Freeman WJ (1996) Reafference and attractors in the olfactory system during odor recognition. Int J Neural Syst 7:489–495.

Kay LM, Laurent G (1999) Odor- and context-dependent modulation of mitral cell activity in behaving rats. Nat Neurosci 2:1003–1009.

Kay LM, Lowry CA, Jacobs HA (2003) Receptor contributions to configural and elemental odor mixture perception. Behav Neurosci 117:1108–1114.

Keller A, Vosshall LB (2004) A psychophysical test of the vibration theory of olfaction. Nat Neurosci 7:337–338.

Kelliher KR, Ziesmann J, Munger SD, Reed RR, Zufall F (2003) Importance of the CNGA4 channel gene for odor discrimination and adaptation in behaving mice. Proc Natl Acad Sci USA 100:4299–4304.

Kesner RP, Gilbert PE, Barua LA (2002) The role of the hippocampus in memory for the temporal order of a sequence of odors. Behav Neurosci 116:286–290.

Kesslak JP, Cotman CW, Chui HC, Van den Noort S (1988) Olfactory tests as possible probes for detecting and monitoring Alzheimer's disease. Neurobiol Aging 9:399–403.

Kirk-Smith M, Booth DA (1980) Effects of androstenone on choice of location in others' presence. In: Olfaction and taste VII, ed. van der Starre H. London: IRL.

Knasko SC (1993) Performance, mood, and health during exposure to intermittent odors. Arch Environ Health 48:305–308.

Knasko SC (1995) Pleasant odors and congruency: Effects on approach behavior. Chem Senses 20:479–487.

Koenderink JJ, van Droom AJ (1979) The internal representation of solid shape with respect to vision. Biol Cybern 32:211–216.

Koester EP, Degel J, Piper D (2002) Proactive and retroactive interference in implicit odor memory. Chem Senses 27:191–206.

Komischke B, Sandoz JC, Lachnit H, Giurfa M (2003) Non-elemental processing in olfactory discrimination tasks needs bilateral input in honeybees. Behav Brain Res 145:135–143.

Koss E, Weiffenbach JM, Haxby JV, Friedland RP (1988) Olfactory detection and identification performance are dissociated in early Alzheimer's disease. Neurology 38:1228–1231.

Kosslyn SM, Ganis G, Thompson WL (2003) Mental imagery: against the nihilistic hypothesis. Trends Cogn Sci 7:109–111.

Kosslyn SM, Thompson WL (2003) When is early visual cortex activated during visual mental imagery. Psychol Bull 129:723–746.

Koster EP (1971) Adaptation and cross-adaptation in olfaction: An experimental study with olfactory stimuli at low levels of intensity. Rotterdam, The Netherlands: Bronder Offset, NV.

Koster EP (2002) The specific characteristics of the sense of smell. In: Olfaction, Taste and Cognition, ed. Rouby C, Schaal B, Dubois D, Gervais R, Holley A, 27–43. Cambridge: Cambridge University Press.

Koster EP, Degel J, Piper D (2002) Proactive and retroactive interference in implicit odor memory. Chem Senses 27:191–206.

Kucharski D, Hall WG (1987) New routes to early memories. Science 238:786–788.

Kuhn TS (1962) The structure of scientific revolutions. Chicago: University of Chicago Press.

Laing DG (1983) Natural sniffing gives optimum odour perception for humans. Perception 12:99–117.

Laing DG, Francis GW (1989) The capacity of humans to identify odors in mixtures. Physiol Behav 46:809–814.

Laing DG, Glemarec A (1992) Selective attention and the perceptual analysis of odor mixtures. Physiol Behav 52:1047–1053.

Laing DG, Legha PK, Jinks AL, Hutchinson I (2003) Relationship between molecular structure, concentration and odor qualities of oxygenated aliphatic molecules. Chem Senses 28:57–69.

Laing DG, Panhuber H, Slotnick BM (1989) Odor masking in the rat. Physiol Behav 45:689–694.

Lansky P, Rospars JP (1993) Coding of odor intensity. BioSystems 31:15–38.

Larjola K, Von Wright J (1976) Memory of odors: Developmental data. Percept Mot Skills 42:1138.

Larson J, Sieprawska D (2002) Automated study of simultaneous-cue olfactory discrimination learning in adult mice. Behav Neurosci 116:588–599.

Larson J, Wong D, Lynch G (1986) Patterned stimulation at the theta frequency is optimal for the induction of hippocampal long-term potentiation. Brain Res 368:347–350.

Larsson M (1997) Semantic factors in episodic recognition of common odors in early and late adulthood: A review. Chem Senses 22:623–633.

Larsson M, Backman L (1997) Age-related differences in episodic odour recognition: The role of access to specific odour names. Memory 5:361–378.

Larsson M, Semb H, Winblad B, Amberla K, Wahlund LO, Backman L (1999) Odor identification in normal aging and early Alzheimer's disease: Effects of retrieval support. Neuropsychology 13:47–53.

Laska M, Alicke T, Hudson R (1996) A study of long-term odor memory in squirrel monkeys (Saimiri sciureus). J Comp Psychol 110:125–130.

Laska M, Hudson R (1993) Discriminating parts from the whole: determinants of odor mixture perception in squirrel monkeys, Saimiri sciureus. J Comp Physiol A 173:249–256.

Laska M, Salazar LT, Luna ER (2003) Successful acquisition of an olfactory discrimination paradigm by spider monkeys, Ateles geoffroyi. Physiol Behav 78:321–329.

Laurent G (1996) Dynamical representation of odors by oscillating and evolving neural assemblies. Trends Neurosci 19:489–496.

Laurent G, Stopfer M, Friedrich RW, Rabinovich MI, Volkovskii A, Abarbanel HD (2001) Odor encoding as an active, dynamical process: experiments, computation, and theory. Annu Rev Neurosci 24:263–297.

Lawless HT (1978) Recognition of common odors, pictures, and simple shapes. Percept Psychophys 24:493–495.

——. (1984) Flavor description of white wine by "expert" and nonexpert wine consumers. J Food Sci 49:120–123.

——. (1996) Flavor. In: Handbook of perception, vol. 16, Cognitive ecology, ed. Friedman MP, Carterrette EC. San Diego: Academic Press.

Lawless HT, Cain WS (1975) Recognition memory for odors. Chem Senses Flavour 1:331–337.

Lawless HT, Engen T (1977) Associations to odors: Interference, mnemonics, and verbal labeling. J Exp Psychol Hum Learn Mem 3:52–59.

Lawson C, Stevenson RJ, Coltheart M (In preparation) Abnormal experience of disgust in Huntingtons disease.

Lehrner PJ, Walla P (2002) Development of odor naming and odor memory from childhood to young adulthood. In: Olfaction, taste, and cognition, ed. Rouby C, Schall B, Dubois D, Gervais R, Holley A, 278–289. Cambridge: Cambridge University Press.

Lehrner PJ, Walla P, Laska M, Deecke L (1999) Different forms of human odor memory: a developmental study. Neurosci Lett 272:17–20.

Leinders-Zufall T, Lane AP, Puche AC, Ma W, Novotny MV, Shipley MT, Zufall F (2000) Ultrasensitive pheromone detection by mammalian vomeronasal neurons. Nature 405:792–796.

Leon M (1992) The neurobiology of filial learning. Annu Rev Psychol 43:377–398.

Leon M, Johnson BA (2003) Olfactory coding in the mammalian olfactory bulb. Brain Res Brain Res Rev 42:23–32.

Levy LM, Henkin RI, Lin CS, Hutter A, Schellinger D (1999) Odor memory induces brain activation as measured by functional MRI. J Comput Assist Tomogr 23:487–498.

Linn CE, Campbell MG, Roelofs WL (1987) Pheromone components and active spaces: What do moths smell and where to they smell it? Science 237:650–652.

Linster C, Cleland TA (2002) Cholinergic modulation of sensory representations in the olfactory bulb. Neural Netw 15:709–717.

Linster C, Hasselmo ME (2000) Neural activity in the horizontal limb of the diagonal band of broca can be modulated by electrical stimulation of the olfactory bulb and cortex in rats. Neurosci Lett 282:157–160.

Linster C, Hasselmo ME (2001) Neuromodulation and the functional dynamics of piriform cortex. Chem Senses 26:585–594.

Linster C, Johnson BA, Yue E, Morse A, Xu Z, Hingco EE, Choi Y, Choi M, Messiha A, Leon M (2001) Perceptual correlates of neural representations evoked by odorant enantiomers. J Neurosci 21:9837–9843.

Linster C, Maloney M, Patil M, Hasselmo ME (2003) Enhanced cholinergic suppression of previously strengthened synapses enables the formation of self-organized representations in olfactory cortex. Neurobiol Learn Mem 80:302–314.

Lipton PA, Alvarez P, Eichenbaum H (1999) Crossmodal associative memory representations in rodent orbitofrontal cortex. Neuron 22:349–359.

Litaudon P, Mouly AM, Sullivan R, Gervais R, Cattarelli M (1997) Learning-induced changes in rat piriform cortex activity mapped using multisite recording with voltage sensitive dye. Eur J Neurosci 9:1593–1602.

Livermore A, Hutson M, Ngo V, Hadjisimos R, Derby CD (1997) Elemental and configural learning and the perception of odorant mixtures by the spiny lobster Panulirus argus. Physiol Behav 62:169–174.

Livermore A, Laing DG (1996) Influence of training and experience on the perception of multicomponent odor mixtures. J Exp Psychol Hum Percept Perform 22:267–277.

Livermore A, Laing DG (1998a) The influence of chemical complexity on the perception of multicomponent odor mixtures. Percept Psychophys 60:650–661.

Livermore A, Laing DG (1998b) The influence of odor type on the discrimination and identification of odorants in multicomponent odor mixtures. Physiol Behav 65:311–320.

Lledo PM, Gheusi G (2003) Olfactory processing in a changing brain. Neuroreport 14:1655–1663.

Lodovichi C, Belluscio L, Katz LC (2003) Functional topography of connections linking mirror-symmetric maps in the mouse olfactory bulb. Neuron 38:265–276.

Logothetis NK, Sheinberg DL (1996) Visual object recognition. Annu Rev Neurosci 19:577–621.

Lorig TS (1999) On the similarity of odor and language perception. Neurosci Biobehav Rev 23:391–398.

Lovibond PF, Shanks DR (2002) The role of awareness in Pavlovian conditioning: Empirical evidence and theoretical implications. J Exp Psychol Anim Behav Process 28:3–26.

Lu XC, Slotnick BM (1994) Recognition of propionic acid vapor after removal of the olfactory bulb area associated with high 2-DG uptake. Brain Res 639:26–32.

Lu XC, Slotnick BM, Silberberg AM (1993) Odor matching and odor memory in the rat. Physiol Behav 53:795–804.

Luo M, Katz LC (2001) Response correlation maps of neurons in the mammalian olfactory bulb. Neuron 32:1165–1179.

Luo M, Katz LC (2004) Encoding pheromonal signals in the mammalian vomeronasal system. Curr Opin Neurobiol 14:428–434.

Lyman BJ, McDaniel MA (1986) Effects of encoding strategy on long-term memory for odours. Q J Exp Psychol 38A:753–765.

Lyman BJ, McDaniel MA (1990) Memory for odors and odor names: Modalities of elaboration and imagery. J Exp Psychol Learn Mem Cogn 16:656–664.

Lynch G (1986) Synapses, circuits and the beginnings of memory. Cambridge, MA: MIT Press.

Maarse H (1991) Volatile compounds in foods and beverages. New York: Marcel Dekker.

MacDonald MK (1922) An experimental study of Henning's system of olfactory qualities. Am J Psychol 33:535–596.

MacLeod K, Laurent G (1996) Distinct mechanisms for synchronization and temporal patterning of odor-encoding neural assemblies. Science 274:976–979.

Mainland JD, Bremner EA, Young N, Johnson BN, Khan RM, Bensafi M, Sobel N

(2002) Olfactory plasticity: one nostril knows what the other learns. Nature 419:802.

Mair R, Capra C, McEntee WJ, Engen T (1980) Odor discrimination and memory in Korsakoff's psychosis. J Exp Psychol Hum Percept Perform 6:445–458.

Mair R, Doty RL, Kelly KM, Wilson CS, Langlais PJ, McEntee WJ, Vollmecke TA (1986) Multimodal sensory discrimination deficits in Korsakoff's psychosis. Neuropsychologia 24:831–839.

Malnic B, Hirono J, Sato T, Buck LB (1999) Combinatorial receptor codes for odors. Cell 96:713–723.

Mann NM (2002) Management of smell and taste problems. Cleve Clin J Med 69:329–336.

Marr D (1982) Vision: A computational investigation into human representation and processing of visual information. San Francisco: Freeman.

Marr D, Nishihara HK (1978) Representation and recognition of the spatial organization of three-dimensional shapes. Proc R Soc Lond 200:269–294.

Martin C, Gervais R, Hugues E, Messaoudi B, Ravel N (2004) Learning modulation of odor-induced oscillatory responses in the rat olfactory bulb: a correlate of odor recognition? J Neurosci 24:389–397.

Martinez BA, Cain WS, de Wijk RA, Spencer DD (1993) Olfactory functioning before and after temporal lobe resection for intractable seizures. Neuropsychology 7:351–363.

Martzke JS, Kopala LC, Good KP (1997) Olfactory dysfunction in neuropsychiatric disorders: Review and methodological considerations. Biol Psychiatry 42:721–732.

Mateo JM, Johnston RE (2000) Kin recognition and the 'armpit effect': evidence of self-referent phenotype matching. Proc R Soc Lond B Biol Sci 267:695–700.

McAlonan K, Brown VJ (2002) The thalamic reticular nucleus: More than a sensory nucleus? Neuroscientist 8:302–305.

McClintock MK (1971) Menstrual synchrony and suppression. Nature 229:244–245.

Melcher JM, Schooler JW (1996) The misremembrance of wines past: Verbal and perceptual expertise differentially mediate verbal overshadowing of taste memory. J Mem Lang 35:231–245.

Menzel R, Muller U (1996) Learning and memory in honeybees: from behavior to neural substrates. Annu Rev Neurosci 19:379–404.

Meredith M (1986) Vomeronasal organ removal before sexual experience impairs male hamster mating behavior. Physiol Behav 36:737–743.

Mesholam RI, Moberg PJ, Mahr RN, Doty RL (1998) Olfaction in neurodegenerative disease. Arch Neurol 55:84–90.

Midkiff EE, Bernstein IL (1985) Targets of learned food aversions in humans. Physiol Behav 34:839–841.

Miles C, Hodder KI (forthcoming) Serial position effects in recognition memory for odours: A reexamination.

Miller WI (1997) The anatomy of disgust. Cambridge, MA: Harvard University Press.

Miwa T, Furukawa M, Tsukatani T, Costanzo RM, DiNardo LJ, Reiter ER (2001) Impact of olfactory impairment on quality of life and disability. Arch Otolaryngol Head Neck Surg 127:497–503.

Moberg PJ, Pearlson GD, Speedie LJ, Lipsey JR, Strauss ME, Folstein SE (1987) Olfactory recognition: Differential impairments in early and late Huntington's and Alzheimer's diseases. J Clin Exp Neuropsychol 9:650–664.

Mombaerts P (1999) Seven-transmembrane proteins as odorant and chemosensory receptors. Science 286:707–711.

———. (2001) How smell develops. Nat Neurosci 4 Suppl:1192–1198.

Moncrieff RW (1951) The chemical senses. London: Leonard Hill.

———. (1966) Odour preferences. London: Leonard Hill.

Moore FW (1970) Food Habits in non-industrial societies. In: Dimensions of nutrition, ed. Dupont J, 181–221. Denver, CO: Associated University Press.

Morgan CD, Murphy C (2002) Olfactory event-related potentials in Alzheimer's disease. J Int Neuropsychol Soc 8:753–763.

Morgan CD, Nordin S, Murphy C (1995) Odor identification as an early marker for Alzheimer's disease: Impact of lexical functioning and detection sensitivity. J Clin Exp Neuropsychol 17:793–803.

Morrot G, Brochet F, Dubourdieu D (2001) The color of odors. Brain Lang 79:309–320.

Moskowitz HR, Barbe CD (1977) Profiling of odor components and their mixtures. Sens Process 1:212–226.

Mozell MM, Kent PF, Murphy SJ (1991) The effect of flow rate upon the magnitude of the olfactory response differs for different odorants. Chem Senses 16:631–649.

Muller D, Gerber B, Hellstern F, Hammer M, Menzel R (2000) Sensory preconditioning in honeybees. J Exp Biol 203:1351–1364.

Murakami M, Kashiwadani H, Kirino Y, Mori K (2005) State-dependent sensory gating in olfactory cortex. Neuron 46:285–296.

Murphy C (2002) Olfactory functional testing: Sensitivity and specificity for Alzheimer's disease. Drug Dev Res 56:123–131.

Murphy C, Cain WS (1980) Taste and olfaction: Independence vs interaction. Physiol Behav 24:601–605.

Murphy C, Cain WS (1986) Odor identification: The blind are better. Physiol Behav 37:177–180.

Murphy C, Cain WS, Gilmore MM, Skinner RB (1991) Sensory and semantic factors in recognition memory for odors and graphic stimuli: Elderly versus young persons. Am J Psychol 104:161–192.

Murphy C, Nordin S, Jinich S (1999) Very early decline in recognition memory for odors in Alzheimer's disease. Aging Neuropsychol Cogn 6:229–240.

Murray JE (2004) The ups and downs of face perception: Evidence for holistic encoding of upright and inverted faces. Perception 33:387–398.

Nagayama S, Takahashi YK, Yoshihara Y, Mori K (2004) Mitral and tufted cells differ in the decoding manner of odor maps in the rat olfactory bulb. J Neurophysiol 91:2532–2540.

Najbauer J, Leon M (1995) Olfactory experience modulated apoptosis in the developing olfactory bulb. Brain Res 674:245–251.

Neath I (1993) Distinctiveness and serial position effects in recognition. Mem Cognit 21:689–698.

Nevitt GA (2000) Olfactory foraging by Antarctic procellariiform seabirds: life at high Reynolds numbers. Biol Bull 198:245–253.

Nevitt GA, Dittman AH (1998) A new model for olfactory imprinting in salmon. Integr Biol 1:215–223.

Nicolaides N (1974) Skin lipids: Their biochemical uniqueness. Science 186:19–26.

Noble AC, Arnold RA, Buechsenstein J, Leach EJ, Schmidt JO, Stern PM (1987) Modification of a standardised system of wine aroma terminology. Am J Enol Viticulture 38:143–146.

Nordin S, Murphy C (1996) Impaired sensory and cognitive olfactory function in questionable Alzheimer's disease. Neuropsychology 10:113–119.

Nordin S, Paulsen JS, Murphy C (1995) Sensory- and memory-mediated olfactory dysfunction in Huntington's disease. J Int Neuropsychol Soc 1:281–290.

Norman JF, Norman HF, Clayton AM, Lianekhammy J, Zielke G (2004) The visual and haptic perception of natural object shape. Percept Psychophys 66:342–351.

Nusser Z, Kay LM, Laurent G, Homanics GE, Mody I (2001) Disruption of GABA(A) receptors on GABAergic interneurons leads to increased oscillatory power in the olfactory bulb network. J Neurophysiol 86:2823–2833.

Ober C, Weitkamp LR, Cox N, Dytch H, Elias S (1997) HLA and mate choice in humans. Am J Hum Genet 61:497–504.

Ohlhoff G, Winter B, Fehr C (1991) Chemical classification and structure-odour relationship. In: Perfumes: Art, science and technology, ed. Lamparsky D, 287–330. London: Elsevier.

Olsson MJ (1999) Implicit testing of odor memory: Instances of positive and negative repetition priming. Chem Senses 24:347–350.

Olsson MJ, Cain WS (1995) Early temporal events in odor identification. Chem Senses 20:753.

Olsson MJ, Faxbrink M, Jonsson FU (2002) Repetition priming in odor memory. In: Olfaction, taste, and cognition, ed. Rouby C, Schaal B, Dubois D, Gervais R, Holley A, 246–260. Cambridge: Cambridge University Press.

Olsson MJ, Friden M (2001) Evidence of odor priming: Edibility judgments are primed differently between the hemispheres. Chem Senses 26:117–123.

Otto T, Eichenbaum H (1992) Complementary roles of the orbital prefrontal cortex and the perirhinal-entorhinal cortices in an odor-guided delayed-nonmatching-to-sample task. Behav Neurosci 106:762–775.

Otto T, Garruto D (1997) Rhinal cortex lesions impair simultaneous olfactory discrimination learning in rats. Behav Neurosci 111:1146–1150.

Owen DH, Machamer PK (1979) Bias-free improvement in wine discrimination. Perception 8:199–209.

Pager J (1974) A selective modulation of the olfactory bulb electrical activity in relation to the learning of palatability in hungry and satiated rats. Physiol Behav 12:189–195.

Pager J (1978) Ascending olfactory information and centrifugal influxes contributing to a nutritional modulation of the rat mitral cell responses. Brain Res 140:251–269.

Paivio A (1991) Dual coding theory: Retrospect and current status. Can J Psychol 45:176–206.

Pangborn RM (1975) Cross-cultural aspects of flavour preference. Food Technol 29:34–36.

Parker A, Ngu H, Cassaday HJ (2001) Odour and Proustian memory: Reduction of context-dependent forgetting and multiple forms of memory. Appl Cogn Psychol 15:159–171.

Pastore RE, Crawley EJ, Berens MS, Skelly MA (2003) "Nonparametric" A' and other modern misconceptions about signal detection theory. Psychon Bull Rev 10:556–569.

Payne TL (1986) Mechanisms in insect olfaction. New York: Clarendon Press.

Pearce TC (1997) Computational parallels between the biological olfactory pathway and its analogue "the elctronic nose": Part I. Biological olfaction. Biosystems 41:43–67.

Perl E, Shay U, Hamburger R, Steiner JE (1992) Taste- and odor-reactivity in elderly demented patients. Chem Senses 17:779–794.

Peron RM, Allen GL (1988) Attempts to train novices for beer flavor discrimination: A matter of taste. J Gen Psychol 115:403–418.

Peto E (1935) Contribution to the development of smell feeling. Br J Med Psychol 15:314–320.

Phillips ML, Young AW, Senior C, Brammer M, Andre C, Calder AJ, Bullmore ET, Perret DI, Rowland D, Williams SCR, Gray JA, David AS (1997) A specific neural substrate for perceiving facial expressions of disgust. Nature 389:495–498.

Pierce JD, Halpern B (1996) Orthonasal and reteronasal odorant identification based upon vapor phase input from common substances. Chem Senses 21:529–543.

Pierce JD, Zeng X, Aronov EV, Preti G, Wysocki CJ (1995) Cross-adaptaion of sweaty-smelling 3-methyl-2-hexenoic acid by a structurally-similar, pleasant smelling odorant. Chem Senses 20:401–411.

Pietrini P, Furey ML, Ricciardi E, Gobbini MI, Wu WH, Cohen L, Guazzelli M, Haxby JV (2004) Beyond sensory images: Object-based representation in the human ventral pathway. Proc Natl Acad Sci USA 101:5658–5663.

Pliner P (1982) The effects of mere exposure on liking for edible substances. Appetite 3:283–290.

Pliner P, Pelchat ML (1991) Neophobia in humans and the special status of foods of animal origin. Appetite 124:999–1002.

Pointer S, Bond NW (1998) Context-dependent memory: Colour versus odour. Chem Senses 23:359–362.

Pol HEH, Hijman R, Tulleken CAF, Heeren TJ, Schneider N, van Ree JM (2002) Odor discrimination in patients with frontal lobe damage and Korsakoff's syndrome. Neuropsychologia 40:888–891.

Polak EH (1973) Multiple profile-multiple receptor site model for vertebrate olfaction. J Theor Biol 40:469–484.

Pollien P, Ott A, Montigon F, Baumgartner M, Munoz-Box R, Chaintreau A (1997) Hyphenated headspace-gas chromatography-sniffing technique: Screening of impact odorants and quantitative aromagram comparisons. J Agric Food Chem 45:2630–2637.

Porter RH, Balogh RD, Cernoch JM, Franchi C (1986) Recognition of kin through characteristic body odors. Chem Senses 11:389–395.

Porter RH, Cernoch JM, McLaughlin FJ (1983) Maternal recognition of neonates through olfactory cues. Physiol Behav 30:151–154.

Potter H, Butters N (1980) An assessment of olfactory deficits in patients with damage to prefrontal cortex. Neuropsychologia 18:621–628.

Prescott J (1999) Flavour as a psychological construct: Implications for perceiving and measuring the sensory qualities of foods. Food Qual Preference 10:1–8.

Price JL, Powell TP (1970) An electron-microscopic study of the termination of the afferent fibres to the olfactory bulb from the cerebral hemisphere. J Cell Sci 7:157–187.

Pylyshn Z (2003) Return of the mental image: Are there really pictures in the brain. Trends Cogn Sci 7:113–118.

Qureshy A, Kawashima R, Imran MB, Sugiura M, Goto R, Okada K, Inoue K, Itoh M, Schormann T, Zilles K, Fukuda H (2000) Functional mapping of human brain in olfactory processing: A PET study. Journal of Neurophysiology 84:1656–1666.

Rabin MD (1988) Experience facilitates olfactory quality discrimination. Percept Psychometrics 44:532–540.

Rabin MD, Cain WS (1984) Odor recognition: Familiarity, identifiability, and encoding consistency. J Exp Psychol Learn Mem Cogn 10:316–325.

———. (1989) Attention and learning in the perception of odor mixtures. In: Perception of complex smells and tastes, ed. Laing DG, Cain WS, McBride RL, Ache BW, 173–188. Sydney: Academic Press.

Ramus SJ, Eichenbaum H (2000) Neural correlates of olfactory recognition memory in the rat orbitofrontal cortex. J Neurosci 20:8199–8208.

Rausch R, Serafetinides EA, Crandall PH (1977) Olfactory memory in patients with anterior temporal lobectomy. Cortex 13:445–452.

Ravel N, Chabaud P, Martin C, Gaveau V, Hugues E, Tallon-Baudry C, Bertrand O, Gervais R (2003) Olfactory learning modifies the expression of odour-induced oscillatory responses in the gamma (60–90 Hz) and beta (15–40 Hz) bands in the rat olfactory bulb. Eur J Neurosci 17:350–358.

Ray JP, Price JL (1992) The organization of the thalamocortical connections of the mediodorsal thalamic nucleus in the rat, related to the ventral forebrain-prefrontal cortex topography. J Comp Neurol 323:167–197.

Raynor HA, Epstein LH (2001) Dietary variety, energy regulation, and obesity. Psychol Bull 127:325–341.

Reed P (2000) Serial position effects in recognition memory for odors. J Exp Psychol Learn Mem Cogn 26:411–422.

Reid IC, Morris RG (1992) Smells are no surer: rapid improvement in olfactory discrimination is not due to the acquisition of a learning set. Proc R Soc Lond B Biol Sci 247:137–143.

Ressler KJ, Sullivan SL, Buck LB (1993) A zonal organization of odorant receptor gene expression in olfactory epithelium. Cell 73:597–609.

Rezek DL (1987) Olfactory deficits as a neurologic sign in dementia of the Alzheimer type. Arch Neurol 44:1030–1032.

Rich TJ, Hurst JL (1999) The competing countermarks hypothesis: reliable assessment of competitive ability by potential mates. Anim Behav 58:1027–1037.

Ridley P, Howse PE, Jackson CW (1996) Control of the behavior of leaf-cutting ants by their 'symbiotic' fungus. Experientia 52:631–635.

Rochefort C, Gheusi G, Vincent JD, Lledo PM (2002) Enriched odor exposure increases the number of newborn neurons in the adult olfactory bulb and improves odor memory. J Neurosci 22:2679–2689.

Rodriguez I, Greer CA, Mok MY, Mombaerts P (2000) A putative pheromone receptor gene expressed in human olfactory mucosa. Nat Genet 26:18–19.

Rolls ET, Critchley HD, Treves A (1996) Representation of olfactory information in the primate orbitofrontal cortex. J Neurophysiol 75:1982–1996.

Rolls ET, Kringelbach ML, de Araujo IET (2003) Different representations of pleasant and unpleasant odours in the human brain. Eur J Neurosci 18:695.

Rolls ET, Rolls JH (1997) Olfactory sensory-specific satiety in humans. Physiol Behav 61:461–473.

Rosch E, Mervis CB, Gray W, Johnson D, Boyes-Braem P (1976) Basic objects in natural categories. Cognit Psychol 8:382–429.

Rosenbluth R, Grossman ES, Kaitz M (2000) Performance of early-blind and sighted children on olfactory tasks. Perception 29:101–110.

Rosenkranz JA, Grace AA (2002) Dopamine-mediated modulation of odour-evoked amygdala potentials during pavlovian conditioning. Nature 417:282–287.

Rosselli-Austin L, Altman J (1979) The postnatal development of the main olfactory bulb of the rat. J Dev Physiol 1:295–313.

Rossiter KJ (1996) Structure-odor relationships. Chem Rev 96:3201–3240.

Royet JP, Hudry J, Zald D, Godinot D, Gregoire MC, Lavenne F, Costes N, Holley A (2001) Functional neuroanatomy of different olfactory judgments. Neuroimage 13:506–519.

Royet JP, Plailly J, Delon-Martin C, Kareken DA, Segebarth C (2003) fMRI of emotional responses to odors: influence of hedonic valence and judgment, handedness, and gender. Neuroimage 20:713–728.

Royet JP, Zald D, Versace R, Costes N, Lavenne F, Koenig O, Gervais R (2000) Emotional responses to pleasant and unpleasant olfactory, visual, and auditory stimuli: a positron emission tomography study. J Neurosci 20:7752–7759.

Rozin E, Rozin S, Rozin E (1992) Ethnic cuisine: How to create the authentic flavors of 30 international cuisines. In: Flavour principle cook book, ed. Rozin E. East Rutherford, NJ: Penguin Group (USA).

Rozin P (1978) The use of characteristic flavourings in human culinary practice. In: Flavour: Its chemical behavioural and commercial aspects, 101–127. Boulder, CO: Westview Press.

Rozin P (1982) "Taste-smell confusions" and the duality of the olfactory sense. Percept Psychophys 31:397–401.

Rozin P (1999) The process of moralization. Psychon Sci 10:218–221.

Rozin P, Fallon A, Augustoni-Ziskind M (1985) The child's conception of food: The development of contamination sensitivity to "disgusting" substances. Dev Psychol 21:1075–1079.

Rozin P, Fallon AE (1987) A perspective on disgust. Psychol Rev 94:23–41.

Rozin P, Haidt J, McCauley C (2000) Disgust. In: Handbook of emotions, 2nd ed., ed. Lewis M, Haviland-Jones JM. New York: Guilford Press.

Rozin P, Hammer L, Oster H, Horowitz T, Marmora V (1986) The child's conception of food: Differentiation of categories of rejected substances in the 16 months to 5 year of age range. Appetite 7:141–151.

Rozin P, Wrzesniewski A, Byrnes D (1998) The elusiveness of evaluative conditioning. Learn Motiv 29:397–415.

Rubin DC, Groth E, Goldsmith DJ (1984) Olfactory cueing of autobiographical memory. Am J Psychol 97:493–507.

Rusiniak KW, Palmerino CC, Garcia J (1982) Potentiation of odor by taste in rats: tests of some nonassociative factors. J Comp Physiol Psychol 96:775–780.

Russell MJ (1976) Human olfactory communication. Nature 260:520–522.

Russell MJ, Cummings BJ, Profitt BF, Wysocki CJ, Gilbert AN, Cotman CW (1993) Life span changes in the verbal categorization of odors. J Gerontol Psychol Sci 48:49–53.

Ryan MJ, Phelps SM, Rand AS (2001) How evolutionary history shapes recognition mechanisms. Trends Cogn Sci 5:143–148.

Saar D, Grossman Y, Barkai E (2001) Long-lasting cholinergic modulation underlies rule learning in rats. J Neurosci 21:1385–1392.

Saar D, Barkai E (2003) Long-term modifications in intrinsic neuronal properties and rule learning in rats. Eur J Neurosci 17:2727–2734.

Sandoz JC, Galizia CG, Menzel R (2003) Side-specific olfactory conditioning leads to more specific odor representation between sides but not within sides in the honeybee antennal lobes. Neuroscience 120:1137–1148.

Sandoz JC, Menzel R (2001) Side-specificity of olfactory learning in the honeybee: Generalization between odors and sides. Learn Mem 8:286–294.

Savage R, Combs DR, Pinkston JB, Advokat C, Gouvier WD (2002) The role of temporal lobe and orbitofrontal cortices in olfactory memory function. Arch Clin Neuropsychol 17:305–318.

Savic I (2002) Brain imaging studies of the functional organization of human olfaction. Neuroscientist 8:204–211.

Schaal B, Coureaud G, Langlois D, Ginies C, Semon E, Perrier G (2003) Chemical and behavioural characterization of the rabbit mammary pheromone. Nature 424:68–72.

Schaal B, Soussignan R, Marlier L, Kontar F, Karima IS, Tremblay RE (1997) Variablity and invariants in early odour preferences: Comparitive data from children belonging to three cultures. Chem Senses 22:212.

Schab FR (1990) Odors and the remembrance of things past. J Exp Psychol Learn Mem Cogn 16:648–655.

Schab FR, Crowder RG (1995) Implicit measures of odor memory. In: Memory for odors, ed. Crowder RG, Schab FR, 71–91. Hillsdale, NJ: Lawrence Erlbaum.

Schaefer ML, Young DA, Restrepo D (2001) Olfactory fingerprints for major histocompatibility complex-determined body odors. J Neurosci 21:2481–2487.

Schafe GE, Bernstein IL (1996) Taste aversion learning. In: Why we eat what we eat:

The psychology of eating, ed. Capaldi ED. Washington, DC: American Psychological Association.

Schank JC (2001) Menstrual-cycle synchrony: Problems and new directions for research. J Comp Psychol 115:3–15.

Schellinck HM, Forestell CA, LoLordo VM (2001) A simple and reliable test of olfactory learning and memory in mice. Chem Senses 26:663–672.

Schemper T, Voss S, Cain WS (1981) Odor identification in young and elderly persons: Sensory and cognitive limitations. J Gerontol 36:446–452.

Schifferstein HP (1997) Perceptual and imaginary mixtures in chemosensation. J Exp Psychol Hum Percept Perform 23:278–288.

Schiffman SS (1974) Physiochemical correlates of olfactory quality. Science 185:112–117.

Schiffman SS (1992) Olfaction in aging and medical disorders. In: Science of olfaction, ed. Serby MJ, Chobor KL, 500–525. New York: Springer-Verlag.

Schiffman SS (1993) Perceptions of taste and smell in elderly persons. Crit Rev Food Sci Nutr 33:17–26.

Schiffman SS, Robinson DE, Erickson RP (1977) Multidimensional scaling of odorants: Examination of psychological and physicochemical dimensions. Chem Senses Flavor 2:375–390.

Schneiderman AM, Hildebrand JG, Brennan MM, Tumlinson JH (1986) Trans-sexually grafted antennae alter pheromone-directed behaviour in a moth. Nature 323:801–803.

Schoenbaum G, Eichenbaum H (1995) Information coding in the rodent prefrontal cortex. I. Single-neuron activity in orbitofrontal cortex compared with that in pyriform cortex. J Neurophysiol 74:733–750.

Schoenbaum G, Setlow B, Ramus SJ (2003) A systems approach to orbitofrontal cortex function: recordings in rat orbitofrontal cortex reveal interactions with different learning systems. Behav Brain Res 146:19–29.

Schoenbaum G, Setlow B, Saddoris MP, Gallagher M (2003) Encoding predicted outcome and acquired value in orbitofrontal cortex during cue sampling depends upon input from basolateral amygdala. Neuron 39:855–867.

Scholz AT, Horrall RM, Cooper JC, Hasler AD (1976) Imprinting to chemical cues: The basis for home stream selection in salmon. Science 192:1247–1249.

Schultz HG (1964) A matching-standards method for characterizing odor qualities. Ann N Y Acad Sci 116:517–526.

Schwartz S, Maquet P, Frith C (2002) Neural correlates of perceptual learning: a functional MRI study of visual texture discrimination. Proc Natl Acad Sci USA 99:17137–17142.

Scott JW (1977) A measure of extracellular unit responses to repeated stimulation applied to observations of the time course of olfactory responses. Brain Res 132:247–258.

Scott JW, Harrison TA (1987) The olfactory bulb: anatomy and physiology. In: Neurobiology of taste and smell, ed. Finger TE, Silver WL, 151–178. New York: Wiley.

Scully RE, Galdabini JJ, McNeely BU (1979) Case 44-1979. New England Journal of Medicine 301:987–994.

Segal SJ, Fusella V (1971) Effect of images in six sense modalities on detection of visual signal from noise. Psychon Sci 24:55–56.

Sejnowski TJ (1996) Anatomical specializations of neurons and their contribution to co-incidence detection. In: Coincidence detection in the nervous system, ed. Konnerth A, Tsien RY, Mikoshiba K, Altman J, 151–159. Strasbourg, France: Human Frontier Science Program.

Sekiguchi T, Suzuki H, Yamada A, Kimura T (1999) Aversive conditioning to a compound odor stimulus and its components in a terrestrial mollusc. Zoolog Sci 16:879–883.

Sewards TV, Sewards MA (2002) Innate visual object recognition in vertebrates: some proposed pathways and mechanisms. Comp Biochem Physiol A Mol Integr Physiol 132:861–891.

Shallice T (1988) From neuropsychology to mental structure. Cambridge: Cambridge University Press.

Shepherd GM (2004) The human sense of smell: are we better than we think? PLoS Biol 2:E146.

Shipley MT, Ennis M (1996) Functional organiztion of olfactory system. J Neurobiol 30:123–176.

Sicard G (1990) Receptor selectivity and dimensionality of odours at the stage of the olfactory receptor cells. In: Chemosenory information processing, ed. Schild D, 21–32. Berlin: Springer-Verlag.

Slotnick B (2001) Animal cognition and the rat olfactory system. Trends Cogn Sci 5:216–222.

Slotnick B, Hanford L, Hodos W (2000) Can rats acquire an olfactory learning set? J Exp Psychol Anim Behav Process 26:399–415.

Slotnick BM (1985) Olfactory discrimination in rats with anterior amygdala lesions. Behav Neurosci 99:956–963.

Slotnick BM, Bell GA, Panhuber H, Laing DG (1997) Detection and discrimination of propionic acid after removal of its 2-DG identified major focus in the olfactory bulb: a psychophysical analysis. Brain Res 762:89–96.

Slotnick BM, Graham S, Laing DG, Bell GA (1987) Detection of propionic acid vapor by rats with lesions of olfactory bulb areas associated with high 2-DG uptake. Brain Res 417:343–346.

Slotnick BM, Kaneko N (1981) Role of mediodorsal thalamic nucleus in olfactory discrimination learning in rats. Science 214:91–92.

Slotnick BM, Katz HM (1974) Olfactory learning-set formation in rats. Science 185:796–798.

Slotnick BM, Kufera A, Silberberg AM (1991) Olfactory learning and odor memory in the rat. Physiol Behav 50:555–561.

Smith BH, Menzel R (1989) The use of electromyogram recordings to quantify odorant discrimination in the honey bee, Apis-Mellifera. J Insect Physiol 35:369–375.

Sobel N, Khan RM, Saltman A, Sullivan EV, Gabrieli JD (1999) The world smells different to each nostril. Nature 402:35.

Sobel N, Prabhakaran V, Zhao Z, Desmond JE, Glover GH, Sullivan EV, Gabrieli JED

(1998) Sniffing and smelling: Separate subsystems in the human olfactory cortex. Nature 392:282–286.

Sobel N, Thomason ME, Stappen I, Tanner CM, Tetrud JW, Bower JM, Sullivan EV, Gabrieli JD (2001) An impairment in sniffing contributes to the olfactory impairment in Parkinson's disease. Proc Natl Acad Sci USA 98:4154–4159.

Solomon GEA (1990) Psychology of novice and expert wine talk. Am J Psychol 103:495–517.

Soussignan R, Schaal B, Marlier L, Jiang T (1997) Facial and autonomic responses to biological and artificial olfactory stimuli in human neonates: Reexamining early hedonic discrimination of odors. Physiol Behav 62:745–758.

Soussignan R, Schaal B, Schmit G, Nadel J (1995) Facial responsiveness to odours in normal and pervasively developmentally disordered children. Chem Senses 20:47–59.

Spelke ES (1990) Principles of object perception. Cognit Sci 14:29–56.

Spors H, Grinvald A (2002) Spatio-temporal dynamics of odor representations in the mammalian olfactory bulb. Neuron 34:301–315.

Sprengelmeyer R, Young AW, Sprengelmeyer A, Calder AJ, Rowland D, Perrett D, Homberg V, Lange H (1997) Recognition of facial expressions: Selective impairment of specific emotions in Huntington's disease. Cogn Neuropsychol 14:839–879.

Staubli U, Fraser D, Faraday R, Lynch G (1987) Olfaction and the "data" memory system in rats. Behav Neurosci 101:757–765.

Staubli U, Le TT, Lynch G (1995) Variants of olfactory memory and their dependencies on the hippocampal formation. J Neurosci 15:1162–1171.

Staubli U, Schottler F, Nejat-Bina D (1987) Role of dorsomedial thalamic nucleus and piriform cortex in processing olfactory information. Behav Brain Res 25:117–129.

Steiner JE, Glaser D, Hawilo ME, Berridge KC (2001) Comparitive expression of hedonic impact: affective reactions to taste by human infants and other primates. Neurosci Biobehav Rev 25:53–74.

Stern K, McClintock MK (1998) Regulation of ovulation by human pheromones. Nature 392:177–179.

Stevens JC, Cain WS, Burke RJ (1988) Variability of olfactory thresholds. Chem Senses 13:643–653.

Stevens JC, Cain WS, Demarque A (1990) Memory and identification of simulated odors in elderly and young persons. Bull Psychon Soc 28:293–296.

Stevens JC, Dadarwala AD (1993) Variability of olfactory threshold and its role in assessment of aging. Percept Psychophys 54:296–302.

Stevenson RJ (2001a) The acquisition of odour qualities. Q J Exp Psychol 54A:561–577.

———. (2001b) Is sweetness taste enhancement cognitively impenetrable? Effects of exposure, training and knowledge. Appetite 36:241–242.

———. (2001c) Associative learning and odor quality perception: How sniffing an odor mixture can alter the smell of its parts. Learn Motiv 32:154–177.

———. (2001d) Perceptual learning with odors: Implications for psychological accounts of odor quality perception. Psychon Bull Rev 8:708–712.

Stevenson RJ, Boakes RA (2003) A mnemonic theory of odor perception. Psychol Rev 110:340–364.

Stevenson RJ, Boakes RA (2004) Sweet and sour smells: Learned synesthesia between the senses of taste and smell. In: The handbook of multisensory processes, ed. Clavert G, Spence C, Stein B, 69–83. Cambridge, MA: MIT Press.

Stevenson RJ, Boakes RA, Prescott J (1998) Changes in odor sweetness resulting from implicit learning of simultaneous odor-sweetness association: An example of learned synesthesia. Learn Motiv 29:113–132.

Stevenson RJ, Boakes RA, Wilson JP (2000a) Counter-conditioning following human odor-taste and color-taste learning. Learn Motiv 31:114–127.

———. (2000b) Resistance to extinction of conditioned odor perceptions: Evaluative conditioning is not unique. J Exp Psychol Learn Mem Cogn 26:423–440.

Stevenson RJ, Case TI (2003) Preexposure to the stimulus elements, but not training to detect them, retards human odour-taste learning. Behav Process 61:13–25.

———. (2005a) Olfactory dreams: Phenomenology, relationship to volitional imagery and odor identification. Imagination Cogn Pers. 24:69–90.

———. (2005b) Olfactory imagery: A review. Psychon Bull Rev 12:244–264.

———. (Submitted) Proactive, but not retroactive interference, in human olfactory perceptual learning.

Stevenson RJ, Case TI, Boakes RA (2003) Smelling what *was* there: Acquired olfactory percepts are resistant to further modification. Learn Motiv 34:185–202.

Stevenson RJ, Case TI, Boakes RA (forthcoming) Implicit and explicit tests of odor memory reveal different outcomes following interference. Learn Motiv.

Stevenson RJ, Dempsey RA, Button M (In preparation) olfactory quality and familiarity: The subjective effects of mere exposure to an odor.

Stevenson RJ, Mahmut M, Sundqvist N (Submitted) Discrimination of familiar odors in children and adults.

Stevenson RJ, Prescott J (1997) Judgments of chemosensory mixtures in memory. Acta Psychol 95:195–214.

Stevenson RJ, Prescott J, Boakes RA (1995) The acquisition of taste properties by odors. Learn Motiv 26:433–455.

———. (1999) Confusing tastes and smells: How odours can influence the perception of sweet and sour tastes. Chem Senses 24:627–635.

Stevenson RJ, Repacholi BM (2003) Age related changes in children's hedonic response to male body odor. Dev Psychol 39:670–679.

———. (2005) Does the source of an interpersonal odour affect disgust? A disease risk model and its alternatives. Eur J Soc Psychol 35:375–401.

Stevenson RJ, Sundqvist N, Mahmut M (Submitted) Age-related changes in odour discrimination: testing a mnemonic account of olfactory perception.

Stopfer M, Laurent G (1999) Short-term memory in olfactory network dynamics. Nature 402:664–668.

Stripling JS, Patneau DK (1999) Potentiation of late components in olfactory bulb and piriform cortex requires activation of cortical association fibers. Brain Res 841:27–42.

Sullivan RM, Hall WG (1988) Reinforcers in infancy: classical conditioning using stroking or intra-oral infusions of milk as UCS. Dev Psychobiol 21:215–223.

Sullivan RM, Hofer MA, Brake SC (1986) Olfactory-guided orientation in neonatal rats is enhanced by a conditioned change in behavioral state. Dev Psychobiol 19:615–623.

Sullivan RM, Taborsky-Barba S, Mendoza R, Itano A, Leon M, Cotman CW, Payne TF, Lott I (1991) Olfactory classical conditioning in neonates. Pediatrics 87:511–518.

Sullivan RM, Wilson DA (1994) The locus coeruleus, norepinephrine, and memory in newborns. Brain Res Bull 35:467–472.

Takahashi M (2003) Recognition of odors and identification of sources. Am J Psychol 116:527–542.

Tanaka JW, Farah MJ (2003) The holistic representation of faces. In: Perception of faces, objects and scenes: Analytic and holistic processes, ed. Peterson MA, Rhodes G, 53–74. New York: Oxford University Press.

Teghtsoonian R, Teghtsoonian M (1982) Perceived effort in sniffing: The effects of sniff pressure and resistance. Percept Psychophys 31:324–329.

———. (1984) Testing a perceptual constancy model for odor strength: The effects of sniff pressure and resistance to sniffing. Perception 13:743–752.

Teghtsoonian R, Teghtsoonian M, Berglund B, Berglund U (1978) Invariance of odor strength with sniff vigor: An olfactory analogue to size constancy. J Exp Psychol Hum Percept Perform 4:144–152.

Teyke T, Wang JW, Gelperin A (2000) Lateralized memory storage and crossed inhibition during odor processing by Limax. J Comp Physiol A 186:269–278.

Tinbergen N, Perdeck AC (1950) On the stimulus situation releasing the begging response in the newly hatched herring gull chick (Larus argentatus Pont.). Behaviour 3:1–39.

Todrank J, Heth G, Johnston RE (1999) Kin and individual recognition: Odor signals, social experience, and mechanisms of recognition. In: Advances in chemical signals in vertebrates, ed. Johnston RE, Muller-Schwarze D, Sorensen P, 289–297. New York: Kluwer Academic.

Todrank J, Wysocki CJ, Beauchamp GK (1991) The effects of adaptation on the perception of similar and dissimilar odors. Chem Senses 16:467–482.

Toone BK (1978) Psychomotor seizures, arterio-venous malformation and the olfactory reference syndrome. Acta Psychiatr Scand 58:61–66.

Tovee MJ, Rolls ET, Ramachandran VS (1996) Rapid visual learning in neurones of the primate temporal visual cortex. Neuroreport 7:2757–2760.

Tronel S, Sara SJ (2002) Mapping of olfactory memory circuits: region-specific c-fos activation after odor-reward associative learning or after its retrieval. Learn Mem 9:105–111.

Tucker D (1963) Physical variables in the olfactory stimulation process. J Gen Physiol 6:453–489.

Turin L (1996) A spectroscopic mechanism for primary olfactory reception. Chem Senses 21:773–791.

Uchida N, Mainen ZF (2003) Speed and accuracy of olfactory discrimination in the rat. Nat Neurosci 6:1224–1229.

Ueno Y (1993) Cross-cultural study of odor perception in Sherpa and Japanese people. Chem Senses 18:352–353.

Ullman S (1996) Object recognition. In: High-level vision, 1–12. Cambridge, MA: MIT Press.

Valdenebro MS, Leon-Camacho M, Pablos F, Gonzalez AG, Martin MJ (1999) Determination of the arabica/robusta composition of roasted coffee according to their sterolic content. Analyst 124:999–1002.

Valentincic T, Caprio J (1994) Consummatory feeding behavior to amino acids in intact and anosmic channel catfish Ictalurus punctatus. Physiol Behav 55:857–863.

van der Klaauw NJ, Frank RA (1996) Scaling component intensities of complex stimuli: the influence of response alternatives. Environ Int 22:21–31.

Vickers NJ, Christensen TA, Hildebrand JG (1998) Combinatorial odor discrimination in the brain: attractive and antagonist odor blends are represented in distinct combinations of uniquely identifiable glomeruli. J Comp Neurol 400:35–56.

Walk HA, Johns EE (1984) Interference and facilitation in short-term memory for odors. Percept Psychophys 36:508–514.

Walk RD (1966) Perceptual learning and the discrimination of wines. Psychon Sci 5:57–58.

Wang HW, Wysocki CJ, Gold GH (1993) Induction of olfactory receptor sensitivity in mice. Science 260:998–1000.

Wang L, Chen L, Jacob T (2004) Evidence for peripheral plasticity in human odour response. J Physiol 554:236–244.

Weckstein LN, Patrizio P, Balmaceda JP, Asch S, Branch DW (1991) Human leukocyte antigen compatibility and failure to achieve a viable pregnancy with assisted reproductive technology. Acta Eur Fertil 22:103–107.

Wedekind C, Seebeck T, Bettens F, Paepke AJ (1995) MHC-dependent mate preferences in humans. Proc R Soc Lond 260:245–249.

Weese GD, Phillips JM, Brown VJ (1999) Attentional orienting is impaired by unilateral lesions of the thalamic reticular nucleus in the rat. J Neurosci 19:10135–10139.

Wehner R, Michel B, Antonsen P (1996) Visual navigation in insects: coupling of egocentric and geocentric information. J Exp Biol 199:129–140.

Weller L, Weller A, Roizman S (1999) Human menstrual synchrony in families and among closest friends: Examining the importance of mutual exposure. J Comp Psychol 113:261–268.

Westbrook RF, Duffield TQ, Good AJ, Halligan S, Seth AK, Swinbourne AL (1995) Extinction of within-event learning is contextually controlled and subject to renewal. Q J Exp Psychol B 48:357–375.

White TL (1998) Olfactory memory: The long and short of it. Chem Senses 23:433–441.

White TL, Hornung DE, Kurtz DB, Treisman M, Sheehe P (1998) Phonological and perceptual components of short-term memory for odors. Am J Psychol 111:411–435.

White TL, Treisman M (1997) A comparison of the encoding of content and order in olfactory memory and in memory for visually presented verbal materials. Br J Psychol 88:459–473.

Williams TJ (1922) Extraordinary development of the tactile and olfactory senses. J Am Med Assoc 79:1331–1334.

Wilson DA (1997) Binaral interactions in the rat piriform cortex. J Neurophysiol 78:160–169.

———. (1998) Habituation of odor responses in the rat anterior piriform cortex. J Neurophysiol 79:1425–1440.

———. (2000) Comparison of odor receptive field plasticity in the rat olfactory bulb and anterior piriform cortex. J Neurophysiol 84:3036–3042.

———. (2001a) Receptive fields in the rat piriform cortex. Chem Senses 26:577–584.

———. (2001b) Scopolamine enhances generalization between odor representations in rat olfactory cortex. Learn Mem 8:279–285.

———. (2003) Rapid, experience-induced enhancement in odorant discrimination by anterior piriform cortex neurons. J Neurophysiol 90:65–72.

———. (2004) Odor perception is dynamic: Consequences for interpretation of odor maps. Chem Sense 30:ii105–ii106.

Wilson DA, Best AR, Sullivan RM (2004) Plasticity in the olfactory system: Lessons for the neurobiology of memory. Neuroscientist 10:513–524.

Wilson DA, Fletcher ML, Sullivan RM (2004) Acetylcholine and olfactory perceptual learning. Learn Mem 11:28–34.

Wilson DA, Stevenson RJ (2003) The fundamental role of memory in olfactory perception. Trends Neurosci 26:243–247.

Wilson DA, Sullivan RM (1994) Neurobiology of associative learning in the neonate: Early olfactory learning. Behav Neural Biol 61:1–18.

Wiltrout C, Dogra S, Linster C (2003) Configurational and nonconfigurational interactions between odorants in binary mixtures. Behav Neurosci 117:236–245.

Woo CC, Coopersmith R, Leon M (1987) Localized changes in olfactory bulb morphology associated with early olfactory learning. J Comp Neurol 263:113–125.

Woo CC, Leon M (1991) Increase in a focal population of juxtaglomerular cells in the olfactory bulb associated with early learning. J Comp Neurol 305:49–56.

Wright GA, Smith BH (2004) Different thresholds for detection and discrimination of odors in the honey bee (Apis mellifera). Chem Senses 29:127–135.

Wright RH (1977) Odor and molecular vibration: neural coding of olfactory information. J Theor Biol 64:473–502.

Wyatt TD (2003) Pheromones and animal behaviour: Communication by smell and taste. New York: Cambridge University Press.

Wysocki CJ, Beauchamp GK (1984) Ability to smell androstenone is genetically determined. Proc Natl Acad Sci USA 81:4899–4902.

Wysocki CJ, Dorries KM, Beauchamp GK (1989) Ability to perceive androstenone can be acquired by ostensibly anosmic people. Proc Natl Acad Sci USA 86:7976–7978.

Wysocki CJ, Pierce JD, Gilbert AN (1991) Geographic, cross-cultural, and individual variation in human olfaction. In: Smell and taste in health and disease, ed. Getchell TV, Bartoshuk LM, Doty RL, Snow JB, 287–314. New York: Raven Press.

Wysocki CJ, Preti G (1998) Pheromonal influences. Arch Sex Behav 27:627–629.

Yang J, Ul Quraish A, Murakami K, Ishikawa Y, Takayanagi M, Kakuta S, Kishi K (2004) Quantitative analysis of axon collaterals of single neurons in layer IIa of the piriform cortex of the guinea pig. J Comp Neurol 473:30–42.

Yee KK, Wysocki CJ (2001) Odorant exposure increases olfactory sensitivity: Olfactory epithelium is implicated. Physiol Behav 72:705–711.

Yeomans MR, Mobini S, Elliman TD, Stevenson RJ (Submitted) Conditioned sensory and hedonic changes for odours paired with sweet and salty tastes in humans. Learn Behav.

Yokoi M, Mori K, Nakanishi S (1995) Refinement of odor molecule tuning by dendrodenritic synaptic inhibition in the olfactory bulb. Proc Natl Acad Sci 92:3371–3375.

Young TA, Wilson DA (1999) Frequency-dependent modulation of inhibition in the rat olfactory bulb. Neurosci Lett 276:65–67.

Yu D, Ponomarev A, Davis RL (2004) Altered representation of the spatial code for odors after olfactory classical conditioning; memory trace formation by synaptic recruitment. Neuron 42:437–449.

Yuan Q, Harley CW, Darby-King A, Neve RL, McLean JH (2003) Early odor preference learning in the rat: bidirectional effects of cAMP response element-binding protein (CREB) and mutant CREB support a causal role for phosphorylated CREB. J Neurosci 23:4760–4765.

Zatorre RJ, Jones-Gotman M (1991) Human olfactory discrimination after unilateral frontal or temporal lobectomy. Brain 114:71–84.

Zatorre RJ, Jones-Gotman M, Evans AC, Meyer E (1992) Functional localization and lateralization of human olfactory cortex. Nature 360:339–340.

Zellner DA, Bartoli AM, Eckard R (1991) Influence of color on odor identification and liking ratings. Am J Psychol 104:547–561.

Zellner DA, Kautz MA (1990) Color affects perceived odor intensity. J Exp Psychol Hum Percept Perform 16:391–397.

Zellner DA, Rozin P, Aron M, Kulish C (1983) Conditioned enhancement of human's liking for flavor by pairing with sweetness. Learn Motiv 14:338–350.

Zhang Y, Burk JA, Glode BM, Mair RG (1998) Effects of thalamic and olfactory cortical lesions on continuous olfactory delayed nonmatching-to-sample and olfactory discrimination in rats (Rattus norvegicus). Behav Neurosci 112:39–53.

Zinyuk LE, Datiche F, Cattarelli M (2001) Cell activity in the anterior piriform cortex during an olfactory learning in the rat. Behav Brain Res 124:29–32.

Zou DJ, Greer CA, Firestein S (2002) Expression pattern of alpha CaMKII in the mouse main olfactory bulb. J Comp Neurol 443:226–236.

Zou Z, Horowitz LF, Montmayeur JP, Snapper S, Buck LB (2001) Genetic tracing reveals a stereotyped sensory map in the olfactory cortex. Nature 414:173–179.

Zucco GM (2003) Anomalies in cognition: Olfactory memory. Eur Psychol 8:77–86.

Zufall F, Leinders-Zufall T (2000) The cellular and molecular basis of odor adaptation. Chem Senses 25:473–481.

Index

Page numbers in *italics* denote figures; those followed by "t" denote tables.

Ingestion, olfaction and, 134–37, 137t
Inhalation of odor molecules, 58–59
Intensity of stimulus, 6, 44; for dispersed pheromones, 79; effect on animal odor quality discrimination, 91–93; individual variation in suprathreshold perception of, 67–69; localization of odor source by, 87–91, 87t; odor familiarity and perception of, 68–69, 215; odor quality and, 71–74, 72t; olfactory information processing of, 215; perceptual constancy and, 69–71, 79, 262; plasticity in judgments of, 69; sniffing rate and, 70–71, 215; spatial variations in, 86

Jacob, T., 66
James, W., 2, 4, 8, 153
Jehl, C., 142, 143, 160, 193
Jellinek, J. S., 180
Jinks, A., 181
Johns, E. E., 185, 192, 193
Johnson, B. A., 73
Johnson, B. N., 70
Johnson, F., 154
Johnston, R. E., 82
Jones, F. N., 191, 192
Jones-Gotman, M., 195, 211–12, 212
Jonsson, F. U., 207
Juxtaglomerular neurons, 48, 52; odor learning and survival of, 111–12, 232

Kautz, M. A., 172–73
Kay, L. M., 117
Kent, P. F., 70, 71
Kinesis, 87t
Kinship and mate odors, 82, 84–85, 96–100, 252
Knasko, S. C., 204
Korsakoff syndrome, 21, 168–70, 169, 187
Korsching, S. I., 73
Koster, E. P., 180, 219
Kuhn, T., 243
Kuisma, J. E., 191

Laing, D. G., 70, 73, 177–79, 178–80, 181
Lancet, D., 71–73
Language, 252–53. See also Naming of odors
Larjola, K., 160

Larrson, M., 165, 166, 167
Laska, M., 127
Laurent, G., 112
Lawless, H. T., 144, 185, 191, 196, 197, 203, 219
Learned odor objects, 2, 6–8, 19, 21, 33–34, 107–8, 122; animal odor discrimination learning, 222–24, 223; animal retention of, 126–29; context-specific, 8, 128, 185–86; hedonic learning, 153–57, 259; human retention of, 181–86, 182; meaning of, 95–103; odor-odor learning, 258; odor set learning, 224, 229, 238–39; paired-associate learning in animals, 225, 228, 240; paired-associate learning in humans, 200–203, 217–18; sensory learning, 142–53, 260–63
Lehrner, P. J., 159
Leon, M., 73
Leslie, J. C., 194
Levy, L. M., 213
Liking of odors, 154–57; developmental changes in, 162
Linster, C., 108
Livermore, A., 124, 178–79, 179, 180
Localization of odor source, 87–91, 87t
Locus coeruleus, 119, 231
Long-term memory, 196–200
Long-term potentiation, 232, 233, 234, 235
Lyman, B. J., 194, 199
Lynch, Gary, 121

Machamer, P. K., 152
Magnetic resonance imaging, functional, 28, 29, 71, 213
Mahmut, M., 161
Mainland, J. D., 67, 70
Major histocompatibility complex-derived odors, 85, 97, 99, 138–39
Maller, O., 172
Mate selection: among animals, 82, 84–85, 96–100; among humans, 138–40
Mathai, K. V., 195
McClintock, M. K., 139
McDaniel, M. A., 194, 199
McNeely, B. U., 210
Meaning of odor stimulus, 6–7, 77, 245, 248; acquired, 95–102; to animals, 7, 93–102, 222, 228; fetal development of, 96;

Meaning of odor stimulus (*cont.*)
food, foraging, and homing odors, 85–
86, 100–102; innate, 93–94, 245; kinship
and mate odors, 82, 84–85, 96–100, 252;
novel odors, 101; in sexual behavior, 98–
99
Melcher, J. M., 151, 152
Memory-based olfactory perception, 2, 8–
9, 32–34, 244, 246, 254; animal retention
of, 126–29; in animals, 107–22, 130–32;
biases in, 129–30; constraints on, 123–32,
260; ecological perspective of, 244–46;
evidence for, 246–49; human retention
of, 181–86, 182; multimodal odor perception and coding, 118–21; odor acuity,
108–15, 231–32; odor memory in animals,
221–42; odor memory in humans, 188–
221; odor mixture synthesis and analysis
in animals, 115–18, 123–26, 131; olfactory
cognition in humans, 188–221; redintegration and, 129, 186. *See also* Odor
memory
Menstrual synchrony, 139
N-Methyl-D-aspartate receptors, 59, 234
Migraine, 210
Mitral cells, 49–50, 51, 52, 111, 112, 230–32;
activation of ß-noradrenergic receptors
of, 235; mitral-granule cell synapse, 234,
235
Mixtures. *See* Odor mixtures
Morrot, G., 172
Mother-infant odor recognition: in ani-
mals, 84–85, 97–98; in humans, 139
Motivational variables, 174–75
Mozell, M. M., 70, 71
Multimodal odor perception and coding,
31, 118–21
Multisensory processing of odors, 119–21
Murphy, C., 160, 196
Murphy, S. J., 70, 71
Muscarinic receptor activation, 233, 239
Mushroom bodies, 47, 230, 236

Naming of odors, 252, 253; aging and, 163,
164, 165; in Alzheimer disease, 166; developmental changes in, 159–60, 160;
difficulty of, 219; discriminability and
deficits in, 166; effect on recognition
memory tasks, 198–200, 202; effects on

olfactory perception, 188–89; familiarity
and, 154, 155; gender differences in, 202–
3; tip-of-the-nose phenomenon and, 219
National Geographic Smell survey, 159
Neural circuits of olfactory and visual sys-
tems, 45–58, 46, 253–57
Neural network modeling, 113, 120
Neural plasticity, 2, 9, 24, 51, 67, 112, 115,
175, 237, 239, 240, 241, 255
Neural substrates of odor memory in ani-
mals, 229–41; explicit memory, 239–41;
implicit memory, 230–38; odor set learn-
ing, 238–39
Neuroimaging studies, 28, 29, 71, 171–72,
213, 259
Neuropsychology, 162–71, 175, 259–60
Ngu, H., 204
Noble, A. C., 16
Norepinephrine, 56, 119–20, 231, 235, 238,
241

Object perception across sensory systems,
22–31; adaptive advantage of, 31–32; auditory, 30–31; cognitive benefits of, 23;
factors affecting, 23; haptic, 29–30; multimodal, 31, 118–21; principles of, 31–32;
temporal coherence and, 23; visual, 24–
29
Object recognition-processing mode, 22–
31, 175; comparing visual and olfactory
information processing, 249–53; ecological perspective of, 244–46; evidence for,
246–49; experimental psychology approach to, 257–60; mnemonically mediated odor perception, 244–49; sensory
physiology approach to, 260–63
Octopamine, 235
Ocular reflexes, 41
Odor acuity in animals, experience-dependent, 108–15, 231–32
Odor descriptors, 12–16; Amoore, 15; Brud,
16; Crocker and Henderson, 13–15, 14;
Henning, 12–13, 13; hierarchy of, 247;
limitations of, 15–16
Odor discrimination, 3, 7–8; animal learning tasks for, 222–24, 223; animal retention
of ability for, 126–29; developmental
changes in, 158–62, 160; familiarity-
enhanced, 143, 257; human retention of

ability for, 181–86, 182; paired trials of, 142–43. *See also* Odor quality discrimination

Odor identification, 3

Odor memory in animals, 221–42; delayed-matching-to-(non)sample tasks of, 225–27, 226; discrimination learning and, 222–24, 223; explicit, 222–28, 223, 226, 227, 239–41; findings from studies of, 241–42; implicit, 230–38; neural substrates of, 229–41; for odor sequences, 227, 227–28; odor set learning, 224, 229, 238–39; paired-associate learning, 225, 228, 240; priming and imagery, 228–29; retention of, 126–29; subclasses of, 229

Odor memory in humans, 188–221; delayed-matching-to-sample tasks of, 168–69, 185, 191, 195; imagery and, 189–90, 208–13, 212; long-term memory, 196–200; paired-associate learning, 200–203, 217–18; priming and, 207–8; retention of, 181–86, 182; retrieval cues and, 203–7, 206; short-term memory, 191–96, 192, 217

Odor mixtures, 5–6, 33, 77, 78, 83; animal limitations on identifying parts of, 123–26; binary, animal perception of, 224–25; complexity rating of, 179–80; configural processing of, 115, 117, 124–25; experience-dependent synthesis and analysis of, in animals, 115–18, 123–26, 131; in foods, 136–37, 137t; good- and poor-blending components of, 177; human limitations on identifying parts of, 177–81, 178–80; interaction of components of, 117–18; of pheromones, animal behavior and, 90–92; redintegration and perception of, 129, 186

Odor molecules: cross-adaptation of, 16–17, 233, 246; inhalation of, 58–59

Odor perception, 2, 41–45; context-specific, 8, 128, 185–86; culture and, 157–58, 258; disregard of object component features for, 43–44; effect of active sniffing on, 44–45, 58–59; experience-dependent, 2, 8–9, 32–34; functions in animals, 76–77; functions in humans, 133–40; imagery in, 189–90, 208–13, 212, 228–29; language and, 252, 253; learned, 2, 6–8, 19, 21, 33–34; of mixtures, 5–6, 7, 77, 78,

83; as mnemonically mediated object recognition, 244–49; model of, 3, 9; of novel odors, 20–21; odor intensity and, 6, 44; spatial component of, 44; stimulus and stimulus acquisition in, 41–45, 43; stimulus intensity and quality of, 64–75; thresholds for, 64–67

Odor prism, 12–13, 13

Odor quality discrimination in animals, 76–132; constraints on two-mode system for, 122–30; functions of animal olfaction, 76–77; limitations on identifying parts of mixtures, 123–26; memory-based system of, 107–22, 130–32; odor meaning, 93–102; physicochemically based system of, 103–7, 130–32; redintegration and, 129; retention of ability for, 126–29; sensory ecology, 77–80, 244–46; stimulus detection, 80–82; stimulus intensity determination, 86–93, 87t; stimulus quality discrimination, 82–86, 84; what function implies about process of, 103

Odor quality discrimination in humans, 133–87; aging and, 163–65, 164, 166; in Alzheimer disease, 165–66, 167, 170; in brain-injured persons, 167–68; constraints on, 176–86; developmental changes in, 158–62, 160; in epilepsy, 167–68, 170; functions of human olfaction, 133–40; hedonic odor learning, 153–57; in Huntington disease, 171; in Korsakoff syndrome, 168–70, 169; limitations on identifying parts of mixtures, 177–81, 178–80; motivational variables and, 174–75; neuroimaging studies of, 171–72; neuropsychology data and, 162–71; redintegration and, 186; retention of ability for, 181–86, 182; sensory odor learning, 142–53; top-down processing for, 172–74; what function implies about process of, 140–42

Odor recognition, 3; delayed-matching-to-sample tasks of, 168–69, 185, 191, 195, 225–27, 226; long-term memory and, 196–200; priming and, 207–8, 228–29; short-term memory and, 191–96, 192, 217

Odor set learning, 224, 229, 238–39

Odors/odorants: background, 6, 7, 8, 33, 42, 53; biologically significant, 18, 86, 222,

Odors/odorants (*cont.*)
248; context-specific, 8, 128, 185–86; cost of, 10; descriptions of, 12; factors affecting transmission of, 79, 79t; feature encoding of, 2, 33–34; feature maps of, 43, 48–49, 49, 58; individual variation in detection of, 64–67; intensity of, 6, 44; liking of, 154–57; meaning of, 6–7, 77; mechanisms of animal localization of, 87–91, 87t; modulatory roles of, 4; novel, 20–21; primary, 15–16, 247; receptor activation by, 48, 79 (*see also* Olfactory receptors); as retrieval cues, 203–7, 206; sensitivity to, 60, 64–67, 66, 109, 231; sources of, 4; sour-smelling, 145–47, 146; spatially discrete, 44; sweet-smelling, 143–47, 146, 155–56

Odor space, 159

Odor stimuli, 4, 10, 41, 76; acquisition of, 41–45, 43; detection of (*see* Stimulus detection); discrimination of, 3, 7–8, 82–86; dispersion as patches, 79; induction of olfactomotor reflexes by, 59; perceptual quality and intensity of, 64–75; spatial region for detection of, 79; stimulus-response models of olfaction, 11–22, 32–33; temporal variations in quality of, 79; variations in sampling of, 59–60

Odor-taste learning: in animals, 127–28, 225; evidence for, 247; in humans, 134–37, 137t, 143–47, 146, 155–56, 181–82, 182; resistance to interference in, 155–56, 181–85, 182; retention of, 181–83, 182; sweet-smelling odors, 143–47, 146; taste potentiation of odor aversions, 225

Odor thresholds, 64–67; age effects on, 163–64; in Alzheimer disease, 165

Olfaction, 4–8; chemosensitivity for, 5; compared with other systems of object recognition, 22–32; functions in animals, 76–77; functions in humans, 133–40; information processing for, 5–7, 19, 190, 213–21, 216; as memory-based object recognition system, 23, 32–34, 244, 246, 254; receptive mechanisms in, 35–63; in single-celled organisms, 102; stimulus-response models of, 11–22

Olfactomotor reflexes, 59

Olfactory bulb, 47–51, 52, 255; animal sati-
ety and responses to food odors, 119; effects of odor learning on interneuron survival in, 111–12; experience-induced changes in synaptic activity in, 233–35; feedback to, 61–63, 118, 256; glomeruli of, 48–49; granule cells of, 48, 50, 111–12, 232, 234; juxtaglomerular neurons of, 48, 111–12, 232; local-field potential oscillations in, 234, 235; mitral and tufted cells of, 49–50, 111, 112, 230–32, 234; oscillatory nature of odor-related activity in, 50–51, 229; perceptual learning and changes in, 232–33

Olfactory cognition, 188–221; imagery, 189–90, 208–13, 212; integrating object recognition process with, 188–90; long-term memory, 196–200; olfactory information processing, 213–21, 216; paired-associate learning, 200–203, 217–18; priming, 207–8; retrieval cues, 203–7, 206; short-term memory, 191–96, 192, 217

Olfactory cortex, 7, 47, 118, 237, 255

Olfactory evoked potentials, 66

Olfactory hallucinations, 209–10

Olfactory imprinting, 81

Olfactory receptor neurons, 47–48, 52

Olfactory receptors, 1, 2, 4, 47, 247; concentration-related recruitment of, 73; differential sensitivities of, 18, 60; discovering types of, 12, 15; genes encoding, 3–4, 6, 42, 47, 59–60; ligands for, 47–48, 117; mammalian repertoire of, 42; mechanisms of odor access to, 58–59; single odorant, 41; specific adaptation of, 16; stimulant intensity and selectivity of, 87

Olfactory retrieval cues, 203–7, 206

Olfactory system, 45–63; circuit anatomy of, 45–58, 46, 253–57; feedback circuits of, 60–63, 118, 256; ontogeny of, 262–63; temporal structure of receptor input to, 59; variation in stimulus sampling and transduction, 58–60

Olfactory tubercle, 47

Olsson, M. J., 207, 208

Ontogeny of olfactory system, 262–63

Orbitofrontal cortex, 56–57, 118, 120, 236, 238, 256; injury to, 168

Owen, D. H., 152

Pager, J., 119
Paired-associate learning: in animals, 225, 228, 240; in humans, 200–203, 217–18
Parker, A., 204
Pearce, T. C., 124
Perceptual constancy, 32, 69–71, 79, 86, 92, 262
Perceptual expertise, 140; effect of exposure and training on, 150–53, 152; in identifying components of odor mixtures, 178–79, 180
Perceptual grouping, 113–15, 114, 117
Perirhinal cortex, 236
Perl, E., 170
Peron, R. M., 152
Pervasive developmental disorder, 170, 175
Peto, E., 161
Petrides, M., 211–12, 212
Petrzilka, M, 65
Pheromones, 5, 6, 41, 66, 73; aggregation, 93; alarm, 78, 81, 84, 93, 94; animal behavior relative to component concentrations in mixtures of, 90–92; ant, 84; contextual effects of, 94; definition of, 105; dispersion as patches, 79; innate meaning of, 94; invertebrate detection of, 81, 105–6; isomers of, 91; for marking territories, 78, 101; mouse, 84; primer, 94; releaser, 94; sexual, 83, 86, 94, 138, 225; two-mode olfactory processing of, 123, 131, 251–52; volatility of, 105
Photoreceptors, 36–37
Physicochemically based olfaction, 11–22, 32–33, 243, 245, 254; in animal models, 103–7, 121, 130–32. *See also* Stimulus-response models of olfaction
Piriform cortex, 47, 51–55, 54, 112–15, 118; autoassociative memory circuits of, 53–55; experience-dependent changes in odor encoding in, 55, 119, 237; extrinsic connections of, 256; in multisensory processing of odors, 119–20; neuronal adaptation to odors, 230–31; odor set learning and changes in, 239; perceptual grouping at, 113–15, 114, 117; perceptual learning effects on, 233; synaptic plasticity in, 53, 237
Plasticity, neural, 2, 9, 24, 51, 67, 112, 115, 175, 237, 239, 240, 241, 255

Plume edge tracking, 89
Polak, E., 3
Porteus Maze, 204
Potter, H., 168
Potts, B. C., 199
Predator odors, 7, 32, 80–81, 93, 245
Prescott, J., 69, 145, 146
Priming: in animals, 228–29; in humans, 207–8, 220

Rabin, M. D., 142, 143, 150, 192–93, 199, 259
Rausch, R., 195
Redintegration: in animals, 129; in humans, 186
Reed, P., 194
Repacholi, B. M., 162
Resistance to interference: in odor memory, 197, 203, 248; in odor-taste learning, 155–56, 181–85, 182
Reticular thalamus, 40–41
Retrieval cues, 203–7, 206
Rheotaxis, 88–89
Roberts, K., 191
Rolls, E. T., 101
Ross, B. M., 191, 192, 196
Ross, D. A., 203
Royet, J. P., 142, 143, 193
Rozin, E., 135
Rozin, P., 155
Russell, M. J., 159

Satiety, 119, 135, 174
Schab, F. R., 204, 207
Schank, J. C., 139
Schemper, T., 165
Schiffman, S. S., 163, 165
Schoenbaum, G., 240
Schooler, J. W., 151, 152, 205
Schott, M. P., 65
Scopolamine, 115, 233
Scully, R. E., 210
Seamon, J. G., 196
Semiochemicals, 105
Sensitivity to odors, 60; exposure-mediated enhancement of, 66, 109; individual variation in, 64–67; long-term changes in, 231
Sensory ecology, 77–80, 244–46

cessing, 249–53; configural face recognition, 25–27, 26; effects of expectation on, 29; experience-dependent, 25, 28–29, 107, 121–22; feature encoding for, 28; haptic object perception and, 30; hue discrimination, 168; language and, 252–53; perceptual grouping, 113–15, 114; sign stimuli in, 25, 31; template matching in, 24–25; testing perceptual learning, 27, 28; view-invariant, 28–29

Visual system, 36–41; circuit anatomy of, 253–57; hierarchical processing by, 38–39, 39; lateral and feedback interactions in, 39–41, 62, 256; object discrimination by, 37, 38; ocular reflexes of, 41; parallel information streams in, 37–38; photoreceptors of, 36–37; spatial discrimination by, 36–37; wavelength and intensity discrimination by, 36

von Clef, J., 173

Von Wright, J., 160
Voss, S., 165

Walk, H. A., 185, 192, 193
Walk, R. D., 152
Walla, P., 159
Wang, L., 66
Waskett, L., 205–6, 206
Westbrook, R. F., 127
White, T. L., 193
Williams, T. J., 153
Wilson, D. A., 108
Wrzesniewski, A., 155, 157

Yeomans, M. R., 155, 156

Zanuttini, L., 160
Zatorre, R., 195, 211–12, 212
Zellner, D. A., 155, 172–73
Zucco, G. M., 197–98